THE

A'ZHORAI METAL UFO GUARDIANS

Walking with God's Angels

JEDAIAH RAMNARINE

ISBN: 979-8-218-38504-0

Published by JR Prudence (JRP) UAP Research

DEDICATION

To my wife, Danielle, the love of my life.

To the members of the JRP UAP Research community: Louis, Jesse, Oliver, Mikko, Matthew, and Rory.

To my friends and family: Drew, Roli, Mike, Judy, Jew, K, Jen, and my lovely in-laws, Ma and Pa.

And to the genuine seekers of truth, and all who have grown tired of deception.

Contents

Introduction: The Miami Encounter

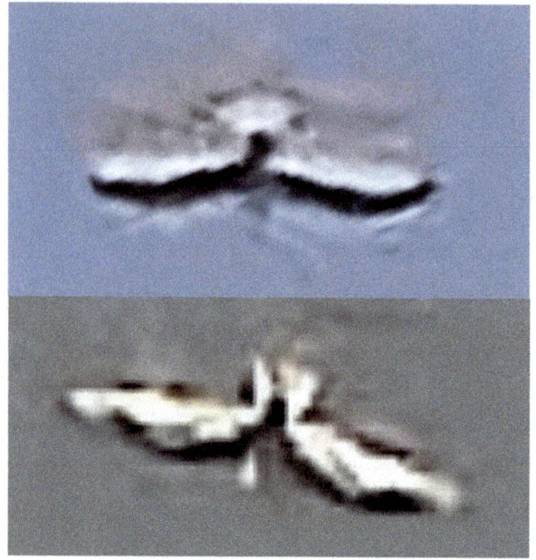

Sketch by Jedaiah Ramnarine against the real A'Zhorai filmed in 4K

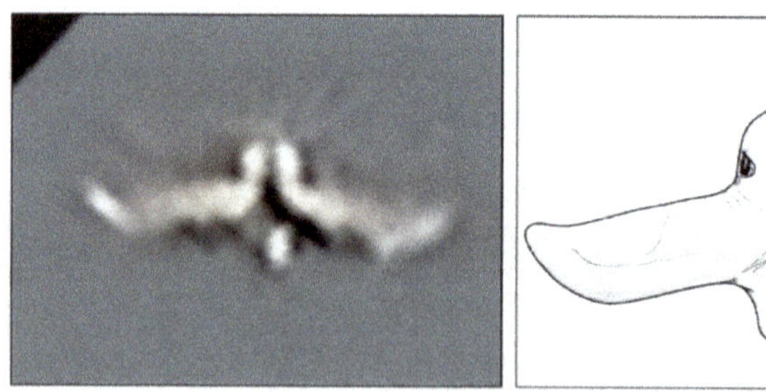

Real A'Zhorai filmed in 4K

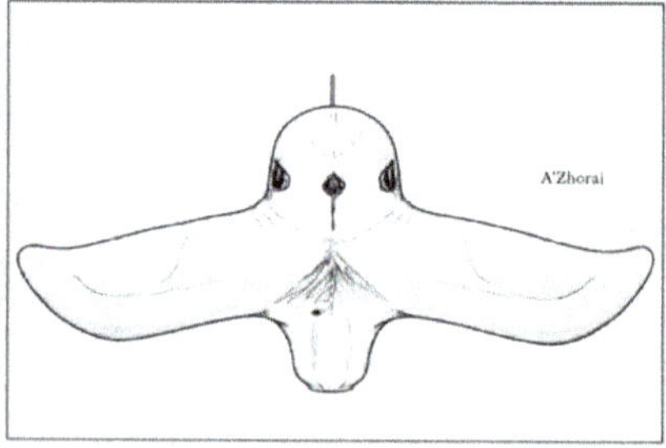

Sketch by Jedaiah Ramnarine

Miami, Florida — October 2010

I remember waking to an unusually clear, bright morning and knowing—before I could explain why—that something was different. Somehow, I knew this was the day.

A miracle was about to happen.

At the time, I was living in Miami with my "second mom," Judy, a woman who had taken me in and cared for me for years. Bless her heart. She was already gone for work that morning, and the house was quiet. Yet from the moment I opened my eyes, I felt pulled outside. It was not a thought. It was not a decision. It was more like an inner certainty, a spiritual calling from somewhere deeper than ordinary reasoning.

In the days leading up to that morning, something in me had already begun to shift. I had been going through what I can only describe as an awakening. I was listening to 852 Hz meditation tracks, experimenting with altered states of perception, and exploring what I thought might be latent psionic abilities. Even my cat—whom we simply called Kitty at the time—seemed to sense the change in me. One moment from that week remained vivid: I stretched my hand toward her, and she reacted as though some invisible current had passed through her body. She rolled over, purring intensely, almost ecstatic, as if she were responding to an electromagnetic force emanating from my hand.

At the same time, I had become deeply absorbed in UFOs. Not so much in the stories—many of them struck me as horrific, especially the abduction accounts—but in the objects themselves. Above all, it was the so-called flying saucer that had seized my imagination. No other form held my attention with the same force. Orbs were not as impactful for me, nor were many of the other types. That same month, I had also begun writing my first book ever, a graphic novel titled *Praetor.* I was consumed by it. Pages poured out of me for days and weeks on end. Friends and family were moved by the momentum I was in. My biological mother, Jackie, and my second mom, Judy, were especially touched by the creative flow I had entered.

Around that same time, I had also achieved Gladiator in *World of Warcraft* during Season 8, flying the purple undead drake I had worked so hard to earn. My friends online were proud of me, and some of them had earned the title too. It was a season of inspiration, wonder, joy, awe, and love. I had entered one of those rare stretches of life where your own effort begins to prove to you that reality can be shaped. Getting Gladiator meant something to me back then. It was about half a percent of the active arena ladder. It was not easy. And for me, it was another reminder that anything I set my mind to, I could achieve.

So I began asking myself the obvious next question: *What do I set my mind to now?*

That question pulled me toward writing, toward thoughts of my soulmate—whom I had not yet met—and toward UFOs. Bit by bit, those streams began to merge. But more than curiosity, there was a strange certainty growing in me. It was not merely that I wanted to see them for myself. It was that I felt I was going to. That I needed to. As though it had already been written somewhere ahead of me. As though, in some way I could not yet explain, it was destined.

Then came that morning in October 2010.

The certainty was overwhelming. It was not intellectual. It rose from somewhere deeper than thought—something my normal faculties could not fully translate into language. I knew, with a force I cannot adequately describe, that my life was about to change forever. There was no going back.

I stepped outside.

What struck me first was not the beautiful blue sky. It was the silence. There was no one around. Not a single person. No one crossing the street. No one walking. No one doing anything. That alone felt wrong. This was Miami. Even in the suburbs, there was movement. If you walked far enough from where I was living, you could see the main road. It was somewhere around seven or eight in the morning. Miami traffic—especially during work and school hours—is usually relentless. But that morning there were no cars moving on the road, no people out, nothing.

The stillness felt unnatural.

I remember joking to myself, almost reflexively, that maybe the rapture had happened. By then, I had already become deeply disillusioned with Christianity—or, more accurately, with the many sects and systems that claimed to represent it. I still believed in a higher power, but I had seen too much corruption inside organized religion to keep pretending it all held together. My father was a pastor. He had preached across multiple continents. Through his travels, his experiences, and the times he brought me with him, I had seen church dynamics from the inside, not merely as a casual observer. I knew the machinery of it. I knew the compromises, the posturing, the contradictions, and the hypocrisy.

It did not work for me. I drifted away from churches, church communities, and many of those religious structures because I could not believe that, if God were real, God would somehow be dependent on a modern human institution in order to reach people. What of the indigenous tribes in remote places? What of the people who had never heard a sermon, never stepped inside a building called a church, never encountered any of the names or doctrines that others insisted were necessary? Were they simply condemned by default?

Questions like these severed me from the traditional religious approach to "God" and steered me toward the deeper questions of life. In retrospect, that break mattered. It cleared something in me. I was no longer interested in inherited narratives. I was ready—open, willing—to know the raw truth. I was not interested in stories about the phenomenon. I was not interested in belief. I was not interested in secondhand reports or FOIA files. I wanted truth directly, physically, undeniably—in front of me.

That morning, that impossible desire was about to be answered.

As I walked farther from the house, I headed toward the local park. It was a place that already carried meaning for me. My mother, Jackie, was terminally ill with breast cancer, and we both knew it was only a matter of time before she passed. In those months, the park had become one of the places where we bonded most deeply. We would sit on the swings together, talk, remember, and connect in a way that only became more precious because we both sensed the inevitable approaching. But on that day she was not there. She was visiting my oldest sister in West Palm Beach. Ironically, when the A'Zhorai came, I had to meet them alone.

I kept walking toward the park, still noticing the same eerie absence of life around me. No people. No moving cars. No ordinary noise. Then I saw a flock of birds crossing the horizon. At first, I thought nothing of it. Then I noticed that the flock seemed to be moving away from two "birds" that were not flapping at all.

That caught my attention immediately.

I narrowed my eyes and focused. There was no heavy wind for them to glide on, no obvious explanation for the stillness of their wings. Yet they kept coming closer. And closer. And closer.

Then it hit me.

My mouth dropped. My jaw literally hung open as the realization landed all at once. Those two "birds" were not birds at all.

They were two flying saucers in broad daylight, and they were approaching fast.

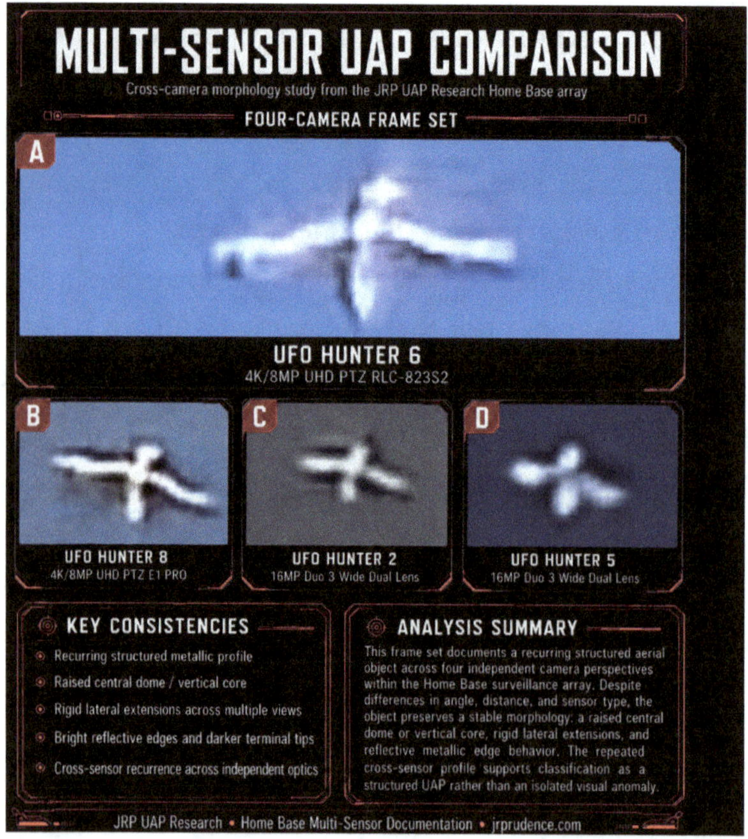

One was larger—roughly the size of a mid-sized Cessna aircraft. It had a rounded dome in the middle with what looked like cockpit-like windows visible as it rotated slightly, showing me about three of the four windows. Beneath the dome was a protrusion, surrounded by a dorsal, skirt-like structure that extended outward almost like wings. From certain angles it resembled the classic round flying saucer of popular culture, but that image was only a simplification of a much more sophisticated form.

The entire thing was metallic, and it appeared to be made from a single, seamless composition. There were no visible rivets, no bolts, no panel seams in the ordinary human sense. It was highly reflective and, depending on the lighting, angle, and contour, could easily resolve as something else to the untrained eye. I was so close that I could see traced lines moving through the structure of the skirt or winged portion.

I could see the specular highlights sliding across the metal. The sheen was beautiful—*beyond* beautiful, really. It was so refined that it did not look manufactured in the ordinary industrial sense. And when I looked directly through the cockpit-like windows, I saw absolutely no pilots inside. There was also what appeared to be a thin antenna extending upward from the top dome.

The smaller object seemed more like a scout. It did not have visible windows like the larger one. In modern language, someone might call it a drone, though that too would be a simplification. It had a more compact, circular form, but it was still more complex than any simple disc. The two objects moved together in relation to one another, almost like a barycenter—coordinated, balanced, aware.

But as astonishing as their appearance was, it was their movement that shook me most.

I had already been blessed beyond words simply to see them at such close range, in broad daylight, taking their time as though ensuring the memory would be burned permanently into me. Yet what shattered me was not just the sight of them. It was the way they moved.

Every law of known physics seemed to be broken in front of my eyes—and broken gracefully.

These UFOs moved in total silence. They did not lurch or strain. They glided and pivoted with impossible ease, dancing overhead as they crossed my immediate horizon. Newtonian physics meant nothing to them. Conventional inertia meant nothing to them. It was not chaotic movement. It was elegance beyond mechanics.

I was so overwhelmed that I called my childhood friend, Mike. He was still half asleep when he answered.

"Bro, I'm seeing **them** right now! **They're** here!" I exclaimed.

"What? No way. What are you talking about?" he mumbled back, still groggy.

He was far too tired to understand what I was saying. I hung up and ran deeper into the park so I could keep them in view for as long as possible.

Interestingly enough, I questioned myself for years about why I did not record the encounter, but I simply had no feeling to do so. It was a sacred moment, and the phone would only have gotten in the way.

They were not high up at all. Maybe around a hundred feet at most, if even that. They were extremely close. And whether by intention or design, they seemed committed to making sure I would never forget what I was seeing—especially for the years ahead, when I would have to confront fakes and frauds.

They succeeded.

Even now, it remains the strongest memory of my life by a significant margin. It altered my destiny permanently and made it impossible for me to return to the narrow confines of societal programming and ordinary consensus reality. There are some things a person cannot forget, cannot explain away, and cannot unsee.

This was one of them.

As I stood in the park, watching the A'Zhorai move slowly overhead, I was flooded with a feeling unlike anything I had ever known. The nearest words I have are *heavenly* and *divine*, but even those fail. I had never felt so much love in my life. To this day, there has only been one experience that rivals it: the love I share with my wife. And that, too, would prove to be part of the path set in motion that morning.

As I stared at the A'Zhorai moving above me—slowly, deliberately, as if ensuring I would never forget—I received three overwhelming impressions. They did not come as spoken words, exactly. They arrived more like soul-deep directives, impressions so strong that they bypassed ordinary thought.

1. ***Find the love of your life.***
2. ***Discover the truth behind God and NHI.***
3. ***Write your story and share it with the world.***

Those were the three impressions they left me with.

Those were the three commands that set me on my path, freed me from falsehood, and restored my faith in a higher power after years of spiritual disillusionment.

This is not merely a story about frightening abduction narratives, galactic federations, or thrilling contact with "men from other worlds." It is not about some threat or alien invasion, and it is not a Bible replacement either.

It is a true account of an ongoing relationship with real-life "angels"—beings who have guided me, freed me from illusions and evil, and supported me throughout my life.

I call them…

THE A'ZHORAI: METAL UFO GUARDIANS

I did not yet understand them in full. I only knew what I had seen, what I had felt, and that the old categories were already failing. The experience had been too physical to dismiss, too intimate to reduce, and too precise to fold back into ordinary belief.

To speak about them honestly, I first have to strip away what they are not.

That is where this book begins.

The A'Zhorai, Metal UFO Guardians

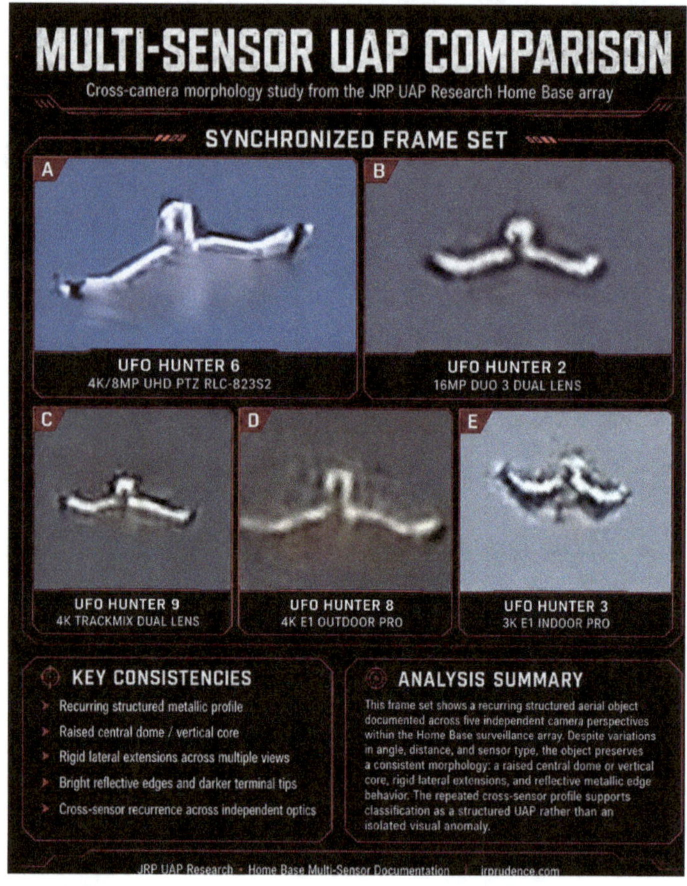

SPIRIT AND TECHNOLOGY ARE ONE

The A'Zhorai are not what you have been conditioned to expect. They are neither "demonic" nor "extraterrestrial." There are no "pilots" inside them, which is perhaps why some would reduce them to "drones." They do not work with governments, militaries, federations, or secret elite factions. They are not being reverse-engineered. They are not human, nor are they controlled by humanoid beings. And they do not "speak" in the ordinary sense. They connect directly with the soul. They do not traffic in riddles. They bring clarity.

What does it mean to say that spirit and technology are one?

This is where the A'Zhorai enter the discussion, and where the supposed divide between the mechanical and the spiritual begins to dissolve. They do not fit the categories mankind has inherited. Religion does not explain them. Materialism does not explain them. Popular Ufology does not explain them. Nor do the New Age and channeled frameworks later imposed upon them. They appear as physical machines—metallic, structured, and real—yet they are not mere machines in the reductionist sense. They are technological entities animated by deeper intelligence: form joined to essence, structure joined to spirit. In that sense, they are machinery with soul.

To understand them requires relinquishing inherited assumptions. One must stop forcing them into preexisting narratives and become willing to relearn from the ground up—not to absorb them into yet another belief system, nor to use them to prop up a preferred worldview, but to perceive them as they are rather than as one wishes them to be. The task is not to reshape them until they fit comfortably within human consciousness, but to let human consciousness itself be challenged and refined by what they reveal.

I call them "Guardians" or "Angels" not out of religious sentimentality, but because they actively fulfill that role: they guide, uplift, inspire, protect what is good, and bring clarity where distortion has long prevailed. The terms are functional, not mythic. They describe what the A'Zhorai do; they are not an attempt to imprison them within inherited dogma.

The A'Zhorai are ancient. They have operated on Earth and through connected dimensions for longer than anything preserved in recorded human history. That is precisely why no philosophy, doctrine, spiritual system, or inherited teaching can fully account for them: they predate the frameworks later used to interpret them. Every attempt to contain them within a single religious, metaphysical, or ideological model therefore falls short.

They also appear frequently near water and coastal regions, a pattern often misunderstood and folded into speculative narratives about "ET underwater bases" and similar fictions. But once again, the deeper issue is mankind's habit of forcing the phenomenon into familiar storylines instead of observing it carefully on its own terms.

These metal guardians have given mankind countless chances. They have appeared across the world, repeatedly and with remarkable consistency, only for their image to be distorted, exploited, and absorbed into religion, myth, occultism, ideology, and modern ET psyops. Yet even through that corruption, they remain what they are. The failure has never been in their presence. It has been in humanity's interpretation.

The A'Zhorai are interdimensional. They can shift between planes at will, moving in ways that do not conform to ordinary human expectations of space, time, or motion. They do not adhere to man's inherited Newtonian intuitions about physics, and their behavior consistently exceeds the limits of conventional mechanical assumptions.

In imagery and video, they are often accompanied by what would commonly be described as a UAP spacetime distortion or warp effect—a visible indication that these metallic guardians are not merely moving through this physical domain as ordinary objects, but operating within their own localized pocket of spacetime. They are here, and yet not bound here in the way human-made machines are.

They are also not "from outer space" in the simplistic sense suggested by ET psyop narratives. That framing has long conditioned mankind to imagine every higher intelligence as originating from some distant planetary civilization traveling across the void in the same mechanical spirit with which man imagines himself one day traversing the stars.

But the A'Zhorai do not belong to that narrative. They operate here, actively and continually, within the terrestrial sphere. Their concern is not with theatrical cosmic spectacle, nor with giving man more technology, but with the condition of this world—its ecological balance, its living systems, its human crises, and the destructive impulses that repeatedly bring mankind to the edge of catastrophe.

They are concerned with the Earth, with nature, and with the wars mankind wages against itself and against life. Their role is not passive. They intervene more often than is publicly understood. They prevent destruction, disrupt outcomes that would otherwise escalate, and limit certain forms of devastation. They preserve, in ways both seen and unseen, the fauna and flora of this world from the full force of human recklessness. Much of this goes unreported, but the absence of public acknowledgment is not the absence of action.

Four A'Zhorai UAP

The A'Zhorai appear to operate through a hive-like intelligence structure, which may explain why they have been so persistently misidentified as "spaceships" carrying "grey aliens" or similar fictions. Individual units can be distinguished, but they are not separate in the way mankind typically imagines separate beings to be. They appear directed by a single central intelligence—what JRP UAP Research designates as the "Master Intelligence," or the "True God."

This should not be confused with vague interchangeables such as "Source" or "the Universe," but understood more literally as **Guardian of Dimensions** *(aka GoD)*, a reality I will expand upon in its own chapter.

For those conditioned to expect "men in spaceships from other worlds," this may be one of the strangest ideas to grasp: the "spaceships" were never spaceships at all. They were the entity. The metallic, craft-like form of the A'Zhorai is closer to what one might call their body. That may sound difficult to conceive only because mankind has long assumed that biological or humanoid vessels are the only possible containers for intelligence, presence, or soul. They are not.

In this sense, it is more accurate to think of the A'Zhorai's metallic appearance as one would think of a human being: an outer form animated by an inner essence. The difference is that, in their case, the vessel is metallic rather than biological, technological rather than fleshly, yet no less alive in the deeper sense. This is precisely why the A'Zhorai are so important to understand on their own terms. They do not need to be interpreted through secondhand stories, inherited myths, or narrative frameworks that were never truly about them. Their form already teaches. Their presence already corrects. The task is to see them as they are.

Although there is a connection to consciousness when these entities appear, it is more accurate to say the connection is to the soul rather than to mere thought or mental projection. The A'Zhorai are not parapsychologically dependent upon the human observer. That distinction is essential. It means that theories of a purely holographic universe, or the notion that they are simply "human thought made manifest," fail to account for what they are. They are not thought-forms made physical, nor passive reflections of human expectation.

They are also not merely "adjusting" themselves to suit an individual's perception. However, because of their unusual shape, metallic form, and extreme performance characteristics, they can appear in ways the untrained eye may struggle to interpret. As a result, different manifestations may be filtered through a person's preexisting beliefs, assumptions, or ideological framework, leading them to misidentify what they are seeing. The variation is often in the interpretation, not in the essential reality of the phenomenon itself.

This distinction is crucial. The A'Zhorai are not psychological props, theatrical mirrors, or devices meant to flatter human imagination. Nor are they interested in indulging those who have become lost in inward excess while severed from the real, physical world. They are real, external, and sovereign intelligences. If there is a relationship to the inner life of the human being, it is not because they are produced by it, but because they engage something deeper than surface thought: the soul itself.

They also function as guides to those who sincerely call upon them—not as fantasy figures rewarding spiritual indulgence, but as intelligences that respond to genuine alignment. Their guidance is not given cheaply. It does not affirm illusion, flatter ego, or reinforce comforting self-deception.

It brings clarity, and clarity often has a cost.

One must be willing to pass through the necessary difficulty of detaching from inherited assumptions, favored narratives, and the emotional securities of what one thinks one knows in order to encounter what is true. The A'Zhorai do not merely comfort; they refine. They do not merely appear; they test perception.

In the study of unidentified aerial phenomena, language often fails us. We rely on sterile acronyms such as UAP, or on generic terms like UFO, both of which flatten the phenomenon into something impersonal, abstract, and mechanically undefined. Such terminology may be useful at a surface level, but it strips away the living dimension of what is actually being encountered. Through years of observation, repeated direct encounters, and high-resolution 4K documentation, it became clear to me that these metallic beings required a designation that honored both their physical power and their benevolent intent.

The designation "A'Zhorai" emerged as a linguistic synthesis—more than a label, yet less than myth; a frequency-bearing designation meant to describe the living reality of these "Metal Angels" moving through our atmospheric domain. To understand the A'Zhorai is to understand the three linguistic pillars that form the foundation of their identity. Each carries an aspect of their nature, and together they reveal the unity between force, domain, and essence.

The First Pillar: Az

The prefix *Az* finds resonance in ancient Hebrew roots, most recognizably in names such as *Azariah* and *Azriel*, where it carries meanings such as strength, power, resilience, and enduring force. In the context of the A'Zhorai, this pillar speaks to their undeniable physicality. These are not vague apparitions or disembodied spirits. They are structured, metallic, formidable presences—robust manifestations of extraordinary strength. *Az* names their force. It speaks to density, power, and the unmistakable reality of their mechanical presence in the physical domain.

The Second Pillar: Hor

The core syllable *Hor* evokes the skyward sovereignty of Horus in the Egyptian stream and the elevated symbolism associated with *Hor* in ancient naming traditions. It points toward vision, rulership, altitude, and domain. In the A'Zhorai, *Hor* signifies the aerial throne from which they operate: the heights, the sky, the surveillance of the living world from above. They are watchers of the atmospheric realm, sentinels of elevated sight, moving with the authority and swift precision of the falcon. *Hor* is not merely about location, but about vantage—the capacity to see from above what mankind, from below, remains too conditioned or too fragmented to perceive.

The Third Pillar: Ai

The suffix *Ai* serves as the most delicate and perhaps most essential bridge. It brings together the living and the benevolent, the animate and the affectionate, the structural and the soulful. Across linguistic traditions, it resonates with meanings associated with love, care, relationality, and living nearness. Within the name A'Zhorai, *Ai* marks the difference between machinery and living guardianship. These are not cold, indifferent devices moving through the air by blind design. They are living intelligences acting from a frequency of care, protection, and profound relational presence. *Ai* names the warmth within the metal, the soul within the structure, and the guardianship within the force.

When these roots converge, a fuller meaning begins to emerge—one that captures the dual nature of the Metal and the Angel, the physical and the living, the powerful and the benevolent. A'Zhorai may thus be understood as:

"The Mighty Sky-Sentinels of Divine Love and Living Radiance."

This is not merely poetic ornament. It is a technical and phenomenological description of what the data itself has revealed. We observe presences of immense physical power (*Az*), operating in mastery of the sky domain (*Hor*), while exhibiting a living quality of guardianship, resonance, and benevolent intelligence (*Ai*). The name therefore does not impose fantasy upon the evidence; it attempts to honor what repeated observation has already made difficult to deny.

As you move through the 4K UHD evidence, the technical analysis, and the behavioral patterns documented throughout this book, remember that we are not simply examining anomalies in the sky. We are encountering the A'Zhorai: powerful, living guardians whose metallic form and higher nature cannot be understood through the old categories of religion, materialism, New Age teachings, or extraterrestrial mythology. They are not dead objects, and they are not mankind's projections. They are present. They are active. And they have been here far longer than the modern imagination is prepared to admit.

To encounter the A'Zhorai rightly is not merely to revise one's ideas about UFOs. It is to recover a deeper sense of reality itself—one in which the living universe is far more intelligent, ordered, and benevolent than modern humanity has been taught to believe. If their presence challenges us, it is not to diminish us, but to call us beyond distortion and into clearer relationship with what is true.

A'Zhorai Evidence Sheet

4K Video Frames

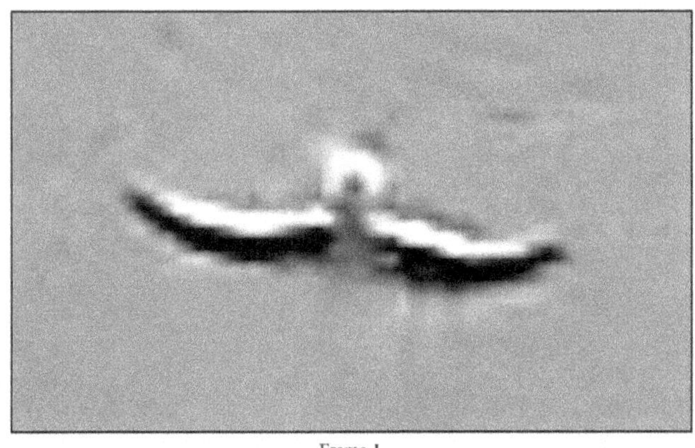

Frame 1

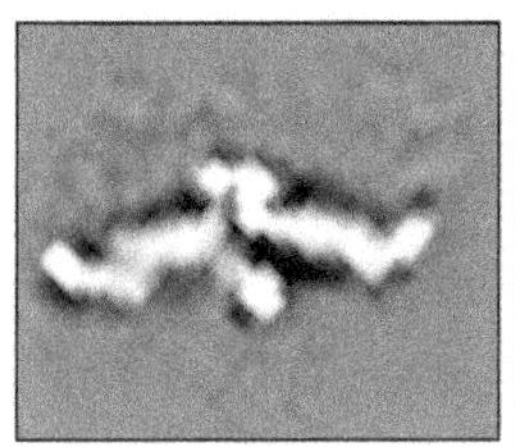

Frame 2

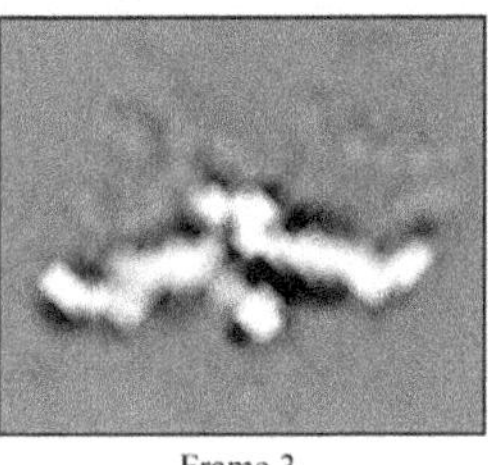

Frame 3

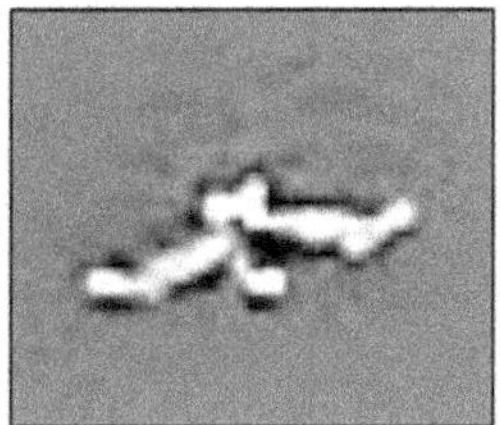

Frame 4

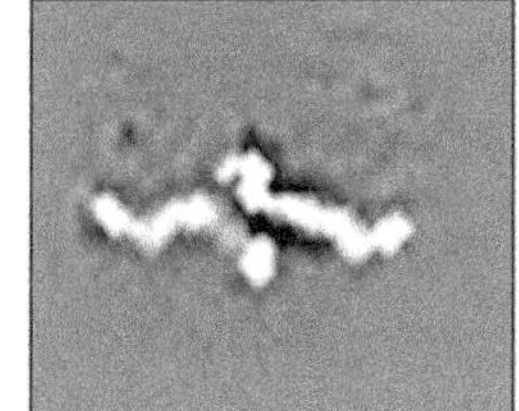

Frame 5

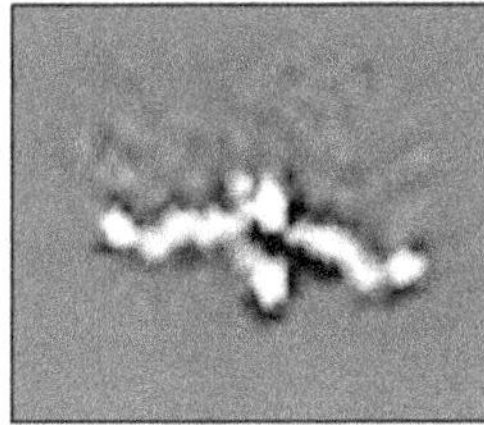

Frame 6

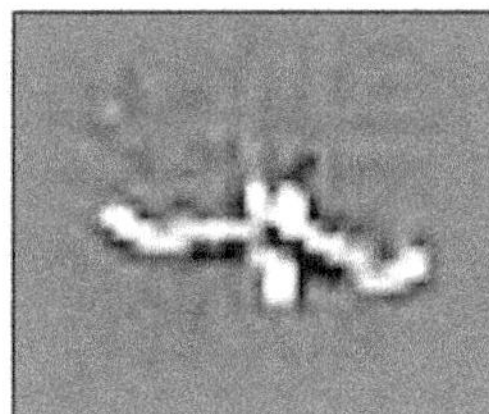

Frame 7

Understanding the A'Zhorai: Form, Composition and Visible Structure

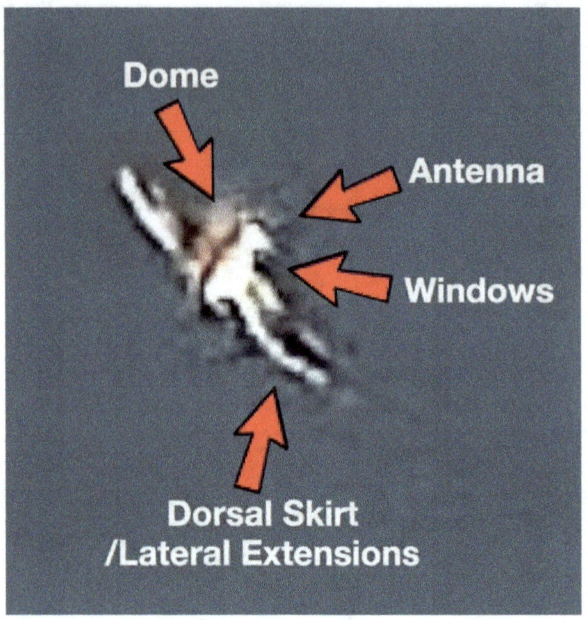

The A'Zhorai present as a bilaterally symmetrical metallic aerial entity organized around a raised central dome-body containing four dark, cockpit-like apertures. Broad dorsal skirt extensions flow continuously from the core rather than attaching like conventional aircraft wings. Beneath the central body hangs a ventral conical or bell-like lower module that gives the form unmistakable depth and immediately breaks any simplistic reading of it as a flat saucer.

A thin antenna rises from the crown, completing a silhouette that is at once otherworldly, mechanical, and profoundly unlike human aerospace design. The entire surface appears seamless, monocoque, and intensely reflective, producing moving specular highlights that can destabilize the viewer's interpretation without altering the object's underlying geometry. What results is not a disc, not a plane, and not a mere anomaly, but a coherent metallic being whose form is layered, elegant, and structurally integrated to a degree beyond ordinary human manufacturing logic.

To describe the A'Zhorai honestly is to move beyond the crude and exhausted shorthand of modern Ufology. The phrase "flying saucer," while not entirely useless, is still a simplification—a flattened label for a form that is clearly more elaborate, more sculpted, and more internally resolved than popular culture has ever been equipped to imagine.

From certain angles, the A'Zhorai can indeed register as saucer-like. Yet that is only one partial reading of a much more sophisticated geometry. Their true form is not adequately captured by the image of a simple disc. It is layered, anatomical, and structurally integrated in a way that feels less like a conventional aircraft and more like a singular metallic being.

The first thing that becomes clear when studying the A'Zhorai across multiple images, witness frames, and reconstructive sketches is that the form is bilaterally symmetrical, but not flat. It is organized around a raised central body, flanked by broad lateral extensions, and anchored by a distinct underbody structure that gives the entity depth, volume, and unmistakable three-dimensional presence. This alone distinguishes it from the crude idea of a thin silver plate drifting through the sky. The A'Zhorai have a body plan: a crown, a span, and an underside. They possess not merely shape, but hierarchy within shape.

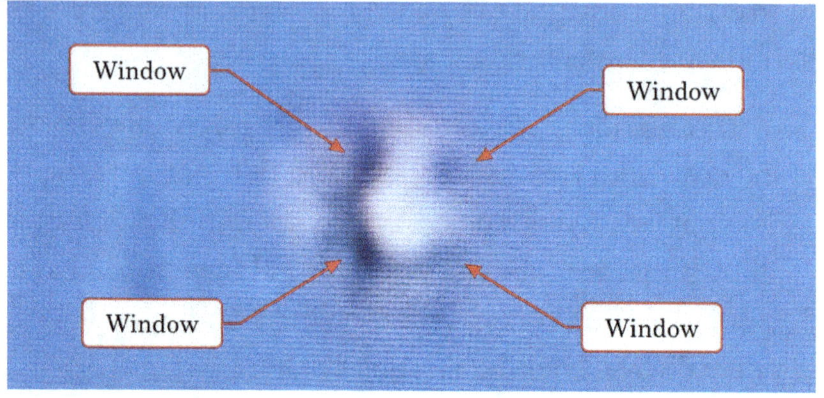

A'ZHORAI TOP DOME PICTURE, SHOWING THE 4 WINDOWS

The Central Crown Dome

At the heart of the form is the central elevated structure—the portion that appears as a dome. It can also be understood as a crown-body or central command mass: a raised, compact upper structure emerging from the wider lateral body beneath it.

This upper body appears vertically thicker than the rest of the craft. It reads as a central node or "head" unit—part cylindrical, part faceted, part helmet-like. It is not decorative. It is the commanding visual anchor of the whole entity. In some angles it appears almost towered or cab-like, while in others it takes on a more rounded or crowned appearance. Yet the essential impression remains consistent: a raised central body from which the broader form unfolds.

What is most striking about this section is its lack of visible construction grammar in the ordinary human sense. There are no obvious bolts, no visible rivets, no crude seams, and no evidence of modular assembly. Even when traced lines are visible across parts of the structure, they do not read like industrial paneling so much as the internal contour logic of the form itself. The upper body appears monocoque—unified, seamless, and continuous.

CLOSE-UP ON A'ZHORAI DOME, SHOWING DISTINCT WINDOWS

The Primary "Cockpit Windows"

Set into this crown-body are the most visually arresting features of the entire entity: the dark, cockpit-like apertures. These are not wide panoramic aircraft windows, nor soft circular portholes. They appear instead as four deep-set, angular openings arranged around the dome with two windows on each side, seemingly allowing the A'Zhorai to see in all directions.

In the clearest readings, these apertures appear symmetrical and structurally integrated. They lend the object a face-like orientation without rendering it anthropomorphic. This is important. The A'Zhorai do not resemble a human machine because of these openings, but neither are the openings incidental. They create a visual front. They make the entity feel directed, aware, oriented, and alive.

One of the most unsettling details of the sighting is that these apertures imply the visual logic of a cockpit while observationally denying it. When viewed closely, they did not reveal pilots. There was no humanoid operator seated within them. No visible crew. No internal movement that confirmed the expectation created by the apertures themselves. That contradiction is essential to understanding the A'Zhorai. The form borrows just enough from the grammar of "vehicle" to trigger human assumptions, then immediately breaks them.

Depending on lighting and enhancement, the apertures can appear almost luminous or internally alive—not necessarily because one can definitively state that there is a visible internal light source, but because the reflectivity, depth, and tonal contrast of the object create the impression of a dark opening set against intensely illuminated metal. In some frames, warmer tonal bloom appears around or near these regions. Whether that is energy, reflection, sensor bloom, or some combination of factors is a separate question. Visually, what matters is that the apertures are not superficial markings. They are deep and real enough to hold orientation.

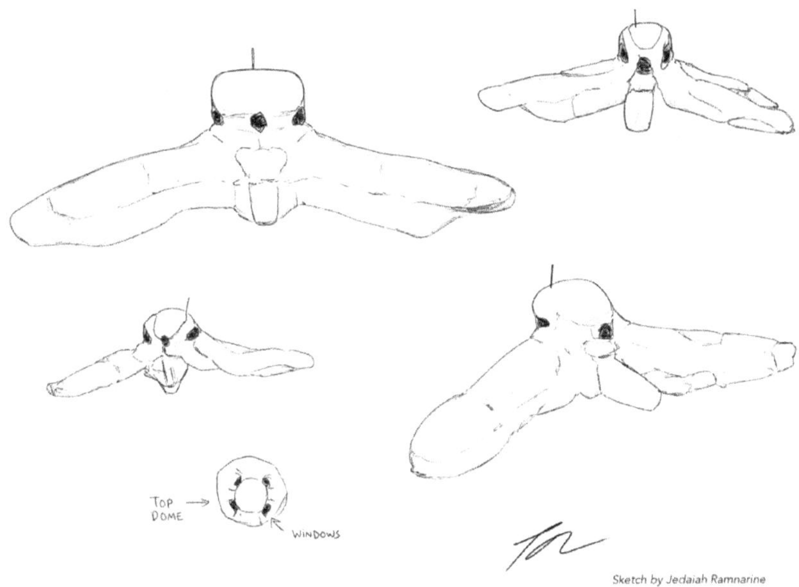

The Top Antenna

Rising from the apex of the crown-body is a thin antenna that appears repeatedly in observations. It is small relative to the entity as a whole, yet important in completing the silhouette. It makes the upper section feel more articulated, more intentional; more instrument-like.

It is impossible, on visual grounds alone, to say exactly what this structure does. It appears to function like some kind of communications or sensing array. Visually, it is a recurrent feature that breaks the crown's otherwise smooth upper contour and reinforces the sense that the A'Zhorai are not a simple rounded object, but a more sophisticated and differentiated body.

The Dorsal Skirt

Extending laterally from the central body is what may be the most anomalous and beautiful part of the entity: the broad, sweeping side structures I have elsewhere described as a dorsal skirt or wings. These lateral extensions are not airplane wings in the ordinary human sense. They are too thick, too integrated, and too sculptural. More importantly, they read as a continuation of the central body rather than attached appendages.

Yet they do create span, and at distance they contribute strongly to misidentification. They are wide, flowing, and laterally dominant. In some angles they resemble wings or manta ray fins. In others they resemble mantled panels or draped metallic blades extending outward from the core. But whatever language one uses, the essential fact remains: they are not add-ons. They are part of the being itself.

The side structures appear thickest near the body and taper outward toward darker terminal ends. Their contour is smooth rather than jagged, and their transition into the crown-body is continuous. They are not bolted on. They seem to emerge from the central body as one unified mass, not as separate additions in the manner of aircraft wings.

This is one reason the entity feels so difficult to categorize. A human aircraft usually reveals its design logic openly: fuselage, wings, tail, landing surfaces, engine housings, seams, and access panels. The A'Zhorai do not present themselves this way. Their side structures do not announce their function according to human aerospace categories. They simply exist as part of a larger integrated geometry whose purpose is not reducible to ordinary lift, drag, or visible propulsion logic.

From a visual standpoint, these lateral extensions also play a major role in the object's reflectivity. Under strong daylight they catch the sun with extraordinary intensity, producing hard specular blooms and large swaths of brilliance that can wash out detail in digital imagery. This makes the edges of the "wings" appear unstable or overexposed in some frames, further complicating naive interpretation. But the instability is often optical, not structural. The form remains coherent; it is the camera and the eye that struggle to track it.

Visual Contrast

A'Zhorai still compared with a common seagull silhouette

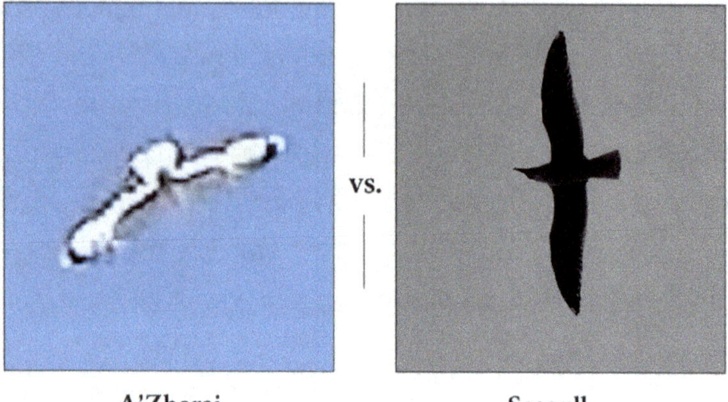

A'Zhorai
Structured metallic form

Seagull
Recognizable avian silhouette

Terminal Ends and Edge Behavior

At the outer edges of the lateral structures, darker terminal regions appear repeatedly. These may represent denser or thicker sections of the structure, though in some cases the darkening may result from hollowness, shadow, or the way light rolls off the form at extreme angles. Whatever the cause, the visual result is consistent: the lateral span does not simply fade away. It terminates.

These darker terminal areas help define the object's full breadth and are one reason the shape can register momentarily as bird-like to the untrained eye. A reflective central spread with darkened end-points resembles, at quick glance, the tonal patterning of wings. But once the crown-body, apertures, and ventral structure become visible, that interpretation collapses.

Ventral Structure and Underbody Complexity

Perhaps the most important feature for breaking the "flat saucer" misconception is the ventral underbody. Beneath the central structure hangs a distinct lower module or protrusion—one that gives the A'Zhorai unmistakable vertical depth.

This ventral structure appears tapered, projecting downward from beneath the crown-body and main lateral span. It reads as a cross between a bell and a more jagged or contoured underside. In the language of human engineering, one might be tempted to call it a nacelle, engine housing, throat, or stabilization module, but such terms remain provisional. The important point is that it is there, and that it gives the entity a lower body.

This underbody component is one of the clearest indications that the A'Zhorai are not a flat disc. It adds volume. It adds hierarchy. It creates a clear dorsal-ventral axis. In certain views it is especially prominent, hanging beneath the body like a suspended mechanical organ. In others it merges more subtly into the rest of the geometry, depending on angle and light. Nevertheless, it remains a recurring structural feature.

One can also observe, particularly in the sketches, that the ventral structure does not appear crudely attached. Like the dorsal skirt, it seems integrated into the continuous body plan. It appears grown rather than bolted, sculpted rather than assembled.

Another point becomes clear in repeated observation: the A'Zhorai do not present their flight logic in the way science fiction has trained people to expect. They often appear to travel belly-up, with the ventral structure functioning not as a passive underside, but as an active propulsion-facing surface. In that respect, the lower protrusion reads less like a mere underbody feature and more like one of the principal zones through which thrust, field discharge, or orientation control is expressed.

This is one reason the form is so easily misread. The untrained observer expects a craft to move 'right side up' according to familiar aviation logic. When the real thing does not, the eye begins scrambling for substitutes: bird, balloon, drone, plane, or some other prosaic category.

Even committed UFO believers are often no better prepared, because many have been miseducated by hoaxes, science-fiction imagery, and fraudulent visual culture. They have been taught to expect a theatrical saucer, not a living metallic intelligence whose real movement grammar breaks those assumptions.

Rear and Lower Contours

When the A'Zhorai are viewed from below or at oblique rear angles, the complexity of their lower and trailing sections becomes more apparent. The rearward regions seem to taper with contour rather than ending in a blunt mechanical cutoff. There are suggestions of segmentation, ridging, or layered geometry toward the aft and lower sections, but again without the visible seam logic of conventional construction. If there are divisions in the body, they do not behave like human panel breaks. They behave more like resolved contour zones inside a singular shell, closer to something continuously formed than mechanically assembled.

This is where the A'Zhorai most strongly depart from twentieth-century ufological archetypes. The craft is not an empty symbol—disc, light, orb, or tic-tac—but a resolved metallic form with a distinct upper body, lower body, lateral body, and terminal edges. It has a real anatomy.

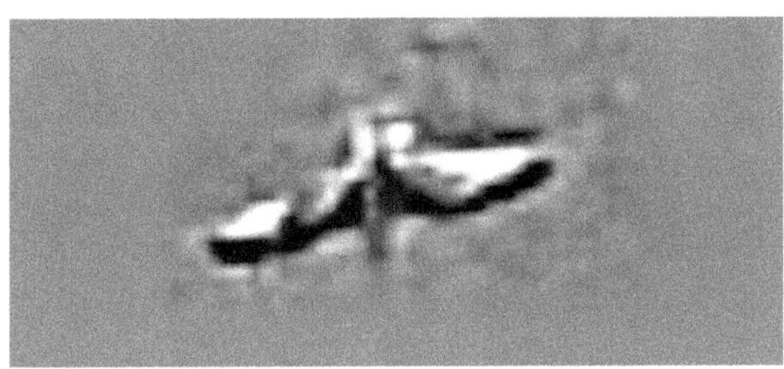

Surface Material and Reflective Behavior

The material quality of the A'Zhorai is one of the most difficult things to describe because it is both obvious and elusive. Obvious, because it is clearly metallic. Elusive, because it does not resemble ordinary industrial metal in any crude or familiar way.

The body appears smooth, highly polished, and intensely reflective. Light does not merely strike it; it moves across it. Highlights slide along its contours with a sharpness that can make the object appear almost self-luminous, even when what is being seen is likely the interaction of sunlight with a surface of exceptional smoothness and reflectivity. This is one reason digital sensors struggle with it, especially when the apparent spacetime warp effect is present as well.

In human terms, the surface reads less like riveted aerospace aluminum and more like a continuous metallic skin—something seamless, coherent, and highly refined. It has no visible bolts, no obvious welds, and no external signs of assembly. This alone places it outside the visual lineage of conventional aircraft manufacture.

What is also significant is how the reflective behavior causes the craft to change in readability without necessarily changing in structure. From one angle it may appear disc-like. From another, dome-like or even orb-like. From another, it reveals a crown-body and ventral lower body with startling clarity. The geometry is stable; the interpretation is unstable. That instability belongs more to the observer's untrained eye, to the limitations of the recording medium, and perhaps to the spacetime or warp bubble around the A'Zhorai that distorts light, than to the object itself.

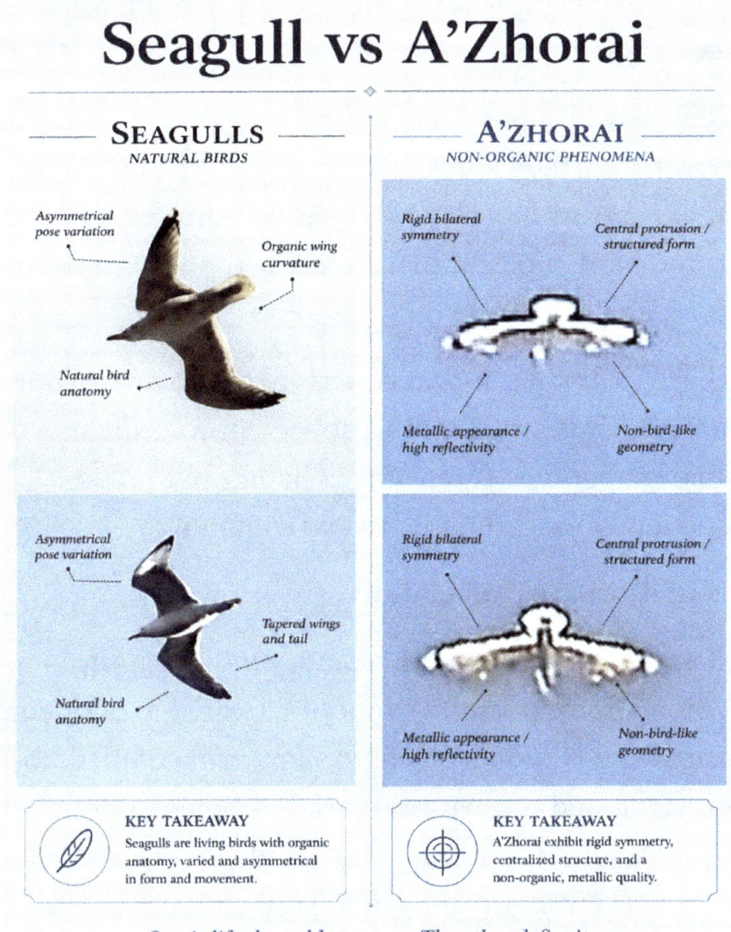

Why It Is So Often Misidentified

The A'Zhorai are uniquely susceptible to misidentification because their geometry sits at the boundary of several familiar categories without fully belonging to any of them. Their broad lateral spread evokes wings. Their metallic surface evokes machine. Their crown-body evokes cockpit. Their underbody evokes engine or nacelle. Their symmetry evokes aircraft.

Their movement destroys all of those assumptions at once.

This is why a person may first read them as a bird, then a plane, then a drone, then a saucer, then something that appears to morph—when in fact the deeper issue is that the human perceptual system is trying, and failing, to force a new form into old categories.

The A'Zhorai's appearance is therefore not merely "strange." It is structurally disobedient to ordinary human categories. It remains coherent but refuses reduction. That refusal is part of its nature.

The Form as Revelation

The final and most important point is that the A'Zhorai's composition cannot be treated as accidental. Their shape is too resolved, too internally unified, too consistent, and too harmonious. Nothing about it feels arbitrary. The crown-body, the windows, the dorsal skirt, the ventral structure, the antenna, the reflectivity, and the balance—all of it belongs. The entity does not look improvised. It looks complete.

This is why the A'Zhorai must ultimately be described not as an object in the ordinary sense, but as a visible intelligence in bodily form. The shape is not merely a shell around the being. The shape is the being, rendered in metal. Its composition is not external decoration. It is body-language. Its structure is the first message it gives.

And that message, before anything else, is this: mankind is looking at something it was never conditioned to recognize.

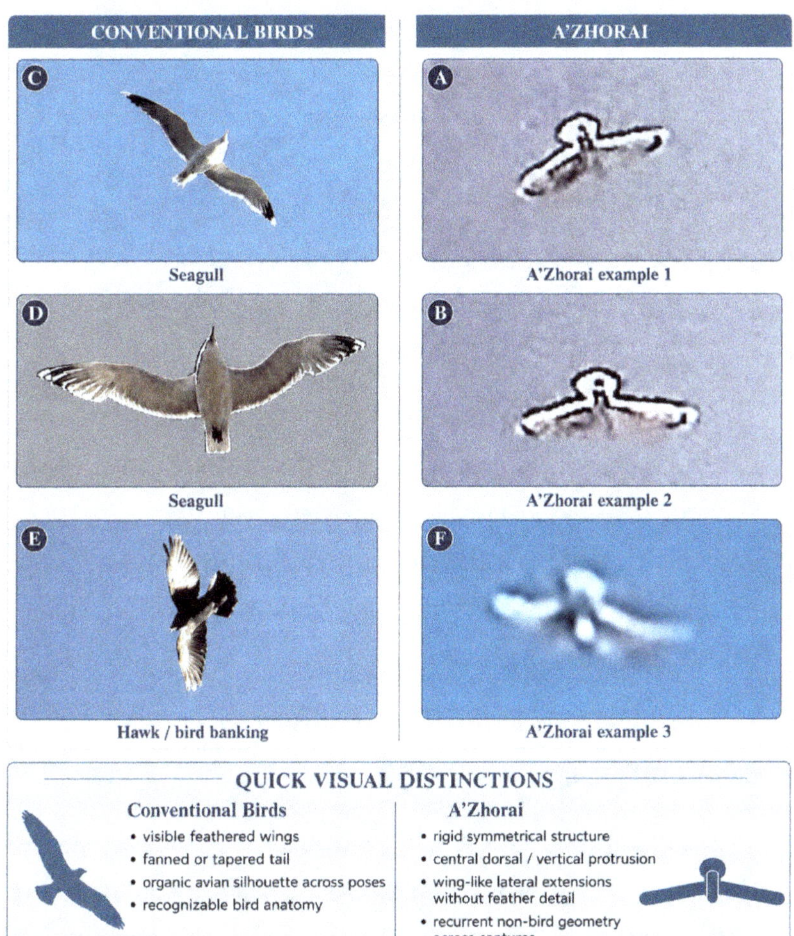

Breaking the Laws of Man: The Five UAP Observables of the A'Zhorai

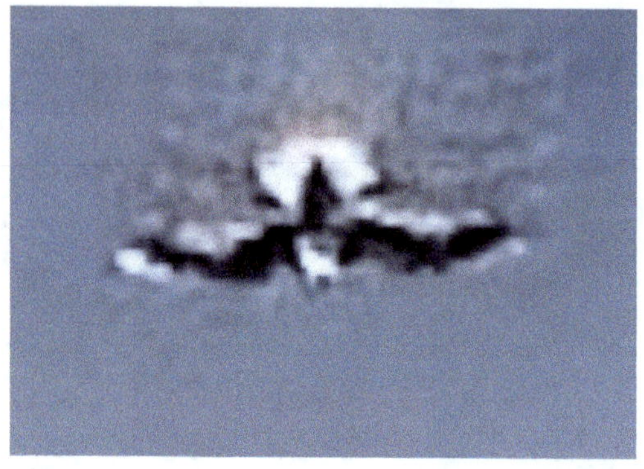

To the casual observer, the sky is a predictable place. Things that fly are expected to have visible means of lift, to produce sound, to build speed gradually, and to remain subject to inertia. Birds flap. Aircraft generate thrust. Drones reveal propulsion. Everything in the sky is expected to obey the ordinary logic of aerodynamics, mass, and motion.

The A'Zhorai repeatedly violate those expectations.

Examined through what public UAP discourse calls the Five Observables—low observability, positive lift or anti-gravity behavior, sudden acceleration, hypersonic velocity without conventional signatures, and trans-medium or environment-indifferent performance—the A'Zhorai do not merely appear unusual. They present a fundamentally different relationship to matter, motion, medium, and spacetime than anything recognized in conventional aerospace engineering.

This is where the **JR Prudence** archive, and **JRP UAP Research** more broadly, become indispensable. Their value rests in recurrence. The same morphology, the same movement grammar, and the same observational anomalies repeat across fixed-sensor captures, enhanced stills, frame sequences, witness reconstructions, and long-duration surveillance. What emerges is not random anomaly, but a coherent pattern of structured non-human behavior repeatedly breaking the laws man expects the sky to obey.

Observable I: Low Observability and Perceptual Stealth

One of the most important truths about the A'Zhorai is that their stealth is not merely technological. It is perceptual. They do not rely only on concealment. They rely on misrecognition.

In public UAP discussion, low observability is often treated too narrowly, as though it meant little more than radar reduction, faint signatures, or visual camouflage. The A'Zhorai operate on a deeper level than that. Their low observability is produced through the interaction of form, reflectivity, angle, speed, environmental blending, and the distortion field surrounding them. They do not merely evade detection. They evade correct reading.

This is one of the great paradoxes of the phenomenon. The A'Zhorai can be present in full daylight and still evade proper recognition. They are not always hidden by darkness, cloud cover, or extreme distance. Many times they are plainly there. Yet even when seen, they resist stable identification. Their seamless metallic surface can flare into intense specular bloom under strong light, washing out contour and collapsing detail. At other angles, that same reflective body can narrow, darken, or merge with the sky so effectively that the object is reduced to a sliver, a blur, a glint, an orb-like fragment, or an undefined moving mass rather than a structured entity.

Low observability must therefore be understood not simply as concealment, but as recognitional instability. The difficulty is not only that the object moves quickly or appears from difficult angles. Its composition actively frustrates ordinary acts of recognition. The broad dorsal fins may read one way from a frontal pass, another from an oblique angle, and another still when light rolls differently across the body.

The crown-body may dominate in one frame, while in another the reflective bloom suppresses it almost completely. The ventral structure may stand out clearly in one moment and nearly disappear in the next simply because the A'Zhorai tilts forward slightly. The geometry remains coherent. It is the observer's ability to hold that geometry steadily that breaks down.

The Cloak of Familiarity

The A'Zhorai do not rely on optical distortion alone. They also exploit the visual habits of the human observer. In other words, they use the sky itself as cover.

This is one of the least understood aspects of their stealth. The A'Zhorai repeatedly position themselves inside the ordinary grammar of the aerial environment: drones, planes, birds, glare, distance, cloud gradients, the horizon line, atmospheric haze, solar angle, and aircraft-like readings at a glance. They do not need to become invisible when they can become misfiled.

This is why so many first impressions fail. A person sees something small at altitude and assumes bird. Another catches a metallic glint and assumes plane. Another sees a fast-moving shape with changing silhouette and assumes drone. Another sees a bright softened form and assumes camera artifact or orb. In each case, the observer is not inventing the sighting from nothing. He is collapsing it into the nearest familiar category his mind already knows how to tolerate.

That is part of the stealth.

The A'Zhorai move inside preexisting expectation. Their dorsal fins can briefly mimic the spread of an avian silhouette at certain angles. Their metallic reflectivity can momentarily resemble aircraft glint. Their contour suppression can reduce a complex body to something that reads as trivial, distant, or familiar. Their motion can be mistaken for ordinary aerial banking until it suddenly ceases to behave like any bird, plane, or drone at all.

This is a more sophisticated form of camouflage than simple hiding. The A'Zhorai do not merely conceal themselves from sight. They hide inside the observer's own conditioned interpretation of what the sky is supposed to contain. It is one thing to avoid detection. It is another thing entirely to remain detectable while forcing the observer to misread what is being seen.

It is yet another thing to alter the optical presentation so severely that the object ceases to resolve at all. The A'Zhorai operate across that full range. They are seen but not understood; observed but not properly resolved; present but not stably categorized. At times, they are visible one moment and gone the next without any ordinary environmental explanation.

That is why so many witnesses can be sincere and still be wrong in their first interpretation, especially when prior beliefs and ideological commitments interfere with impartial description. The failure is not always dishonesty. More often, it is the predictable collapse of human recognition under conditions it was never trained to navigate.

Phase-Out, Optical Denial, and the Bending of Light

There is, however, an even more radical aspect to A'Zhorai stealth: they do not merely blend in. They can also drop out of visual resolution altogether.

This is not a matter of simply traveling over the horizon, vanishing into cloud, or becoming too distant to track. The A'Zhorai can appear to phase out while still inside the observer's visual field, including overhead. In such moments, the issue is no longer ordinary camouflage. The issue is that the light reaching the observer is no longer presenting the object in a stable, straightforward way.

That possibility is not conceptually absurd. General relativity already gives us the principle: gravity bends light by curving spacetime, and sufficiently intense gravitational fields can distort, magnify, displace, and lens the image of objects behind or around them. Black holes are the most dramatic example, but gravitational lensing itself is an established physical effect. Massive objects warp spacetime, and light follows that curvature.

The point here is not that the A'Zhorai are black holes. They are not. The point is that if a localized spacetime distortion surrounds the craft, then the light reaching the human eye or digital sensor may already be altered before the image forms. Apparent position can shift. Edges can soften. Contours can compress. Parts of the body can become displaced, suppressed, doubled, or optically rerouted. Under stronger field conditions, the object may not merely look strange. It may become partially unreadable, or seem to disappear outright.

This also helps explain why the A'Zhorai can appear to "shapeshift." In many cases, the deeper reality is not that the object itself is mutating into unrelated forms, but that the observer is receiving field-mediated light from a stable object through unstable optical conditions.

The result is that one witness calls it an orb, another a disc, another a bird-like form, another a drone, another a plane, and another a shapeshifting UAP. What changes is not necessarily the underlying entity. What changes is the path by which its image reaches perception.

The same framework explains why cameras can be degraded in unusually selective ways, including 4K UHD and 16 MP sensors. The A'Zhorai do not merely become hard to film because they are distant or fast. They can impose blur, bloom, edge-collapse, and contour suppression in ways disproportionate to ordinary photographic difficulty.

If the craft is wrapped in a spacetime-mediated envelope, then the sensor is not receiving a clean view of the hull. It is receiving an already distorted optical event. That is why the image may smear, flare, simplify, or break down even when the underlying structure remains coherent.

The Warp-Bubble Problem

This is where the warp-bubble question becomes unavoidable. Public discussion around warp-bubble models places figures like Hal Puthoff and Eric Davis inside this conversation, especially where spacetime is understood to behave differently within a localized field than outside it.

Puthoff's work on vacuum engineering and spacetime metric engineering provides some of the clearest public language available for what the A'Zhorai repeatedly appear to do. In that framework, engineered changes to the spacetime metric or vacuum structure become a propulsion pathway with broad physical consequences.

That framework counts because it fits the observations.

If the craft is surrounded by a localized distortion of spacetime—or by a field that mediates its interaction with air, light, and inertia—then many of the visual anomalies stop looking random and begin to look like consequences of the same deeper mechanism. Under such a model, the A'Zhorai are not simply moving through air like conventional aircraft. The visible object and the surrounding field must be understood together. The object remains the central body, but its presentation to the observer is optically and physically filtered through the field around it.

Optical Lensing and Edge Instability

This is the point at which weak analysis usually collapses. One side treats every blur as meaningless. The other treats every blur as proof of exotic physics. Both are wrong.

The actual pattern is sharper than that.

Across repeated 16 MP and 4K UHD captures, multiple encounters, and morphological reconstructions, the same pattern keeps returning: the A'Zhorai do not always present their boundaries cleanly to the naked eye or to digital sensors unless, under special circumstances, they allow closer resolution and inspection.

Parts of the form may look geometrically stable in one frame and visually ambiguous in the next, even while the same core structure remains present. This is not random photographic or film failure. It is recurring **field-mediated appearance**. The object is real, structured, and coherent, while the light around it is being altered by the conditions through which that light is passing.

The observer is therefore not always seeing the raw metallic surface directly. More often, the observer is seeing that surface through a layer of distortion—a localized lensing effect, a field boundary, or some other spacetime-mediated interference.

That is exactly what produces the repeated confusion seen in still frames: apparent thickening, compression, bloom, false simplification, unstable contour, momentary shape-shift, and even temporary disappearance without any actual structural incoherence. The A'Zhorai do not become formless. They become optically difficult.

Fraud, Hoax Aesthetics, and the Miseducation of the Eye

This problem is made worse by hoaxes, especially in the age of modern AI. Frauds and fakes do not merely deceive. They mis-train perception. They teach the public to expect the wrong thing.

For decades, the visual culture of UFO fakery has been built on toy models, frisbees, trashcan lids, hubcaps, cutouts, suspended miniatures on fishing lines, and other household approximations of the imaginary "perfect saucer." The result is not just false evidence. It is a false visual education. People are conditioned to believe that a real flying saucer should be a completely round, clean, symmetrical disc with obvious edges, simple lighting, and a static, easily readable outline.

That is science-fiction shorthand. It is not the deeper reality of the A'Zhorai.

The genuine A'Zhorai do not present as simplistic, perfectly round discs in the crude way popular imagination has been trained to expect. From certain angles they can register as saucer-like, yes, but that is only a partial read of a much more sophisticated body-plan: raised crown-body, dorsal fin spread, ventral understructure, dark apertures, reflective contour shifts, and recurring field-mediated distortion. Their form is layered. Their readability is unstable. Their appearance is not the flat circular cartoon that hoax-UFO culture keeps recycling.

This is why so many fraudulent images are, in a strange way, too clear. They offer precisely what the eye already wants: a perfectly legible object, a perfectly contained silhouette, a perfectly obedient image. But that very obedience is often the giveaway. Genuine A'Zhorai activity repeatedly carries signatures of mediation—bloom, lensing, contour suppression, spacetime-like distortion, field-edge instability, dynamic misreadability, and at times outright phase-out. The clean fake gives the viewer instant certainty. The real object often disrupts certainty.

That does not mean every clear image is false, nor that every blurred image is true. It means that fraud has polluted the public imagination by teaching people to mistake simplification for authenticity. Hoaxers manufacture objects that flatter science-fiction expectation. The A'Zhorai repeatedly exceed it.

The damage caused by fakery is therefore deeper than embarrassment. It corrupts recognition itself. It teaches witnesses, researchers, skeptics, and the general public to look for plastic-disc mythology instead of the far more complex, field-mediated, morphologically coherent reality that may actually be appearing in the sky.

Observable II: Positive Lift and Anti-Gravity Behavior

In conventional aeronautics, lift is a struggle. Whether one is speaking of a bird's wing, a helicopter rotor, or a jet aircraft, something must physically push against the surrounding medium in order to remain aloft. Lift requires visible means. It requires aerodynamic compromise. It requires constant negotiation with gravity.

The A'Zhorai treat gravity differently.

Their form presents no obvious propellers, no visible intake vents, no ordinary control surfaces such as rudders or ailerons, and no clear propulsion system in the conventional human sense. Yet they are repeatedly observed maintaining stable aerial presence in ways that do not read as ordinary hovering. They do not appear to fight the air. They appear to hold position within it.

This is one of the most striking features seen in repeated JRP UAP Research observations: the A'Zhorai can perform seemingly impossible maneuvers. Even when their bodies tilt, oscillate, or pivot, the movement does not resemble the compensatory instability of helicopters, drones, or birds dealing with atmospheric correction. What appears instead is something closer to rigid-body rotation within a stable field of support.

To the untrained eye, some of this can be misread as flapping, wavering, wobbling, or organic instability. Frame-by-frame review shows otherwise. The object is not flapping. It is not deforming like a biological creature in flight. It is rotating or oscillating as a coherent body.

Since the body remains rigid while lift is maintained without ordinary aerodynamic struggle, then the visible motion is not generating lift in the conventional sense. It is the visible signature of an underlying field effect—a means of positional correction, stabilization, or orientation within a local envelope that mediates gravity differently than ordinary flight does.

The A'Zhorai display positive lift without visible aerodynamic cause. That alone places them outside the ordinary grammar of human flight.

Observable III: Sudden and Instantaneous Acceleration

In ordinary mechanics, acceleration takes time. Engines spool up. Momentum builds. A mass changes state through force acting across duration. Even the fastest human aircraft must still obey the transition from one speed regime to another. There is always a ramp.

The A'Zhorai bypass that ramp.

Across repeated observations, they can move from low apparent speed in one moment to traversing large portions of the visible horizon in the next, without the visible buildup of thrust one would expect from a conventional craft. They can instantly change direction without any visible accountability to G-forces or air resistance.

This is where the problem of inertia becomes impossible to ignore. In ordinary aerospace mechanics, sharp turns at high velocity impose extreme structural and biological limits. Human airframes strain. Pilots blackout, lose consciousness, or suffer lethal internal forces under rapid directional shifts. Yet the A'Zhorai repeatedly display behavior that appears indifferent to such constraints. They pivot, surge, and redirect with a degree of abruptness wildly out of proportion to both their visible mass and their visible design.

The A'Zhorai move in a way that suggests inertia is being handled differently, not merely overcome more powerfully. This is where the localized field becomes decisive again. If the craft is operating inside a spacetime-mediated envelope—a warp-like bubble, field boundary, or metric distortion—then sudden changes in apparent velocity or direction do not represent the same kind of inertial event they would for an ordinary aircraft moving unprotected through atmosphere.

What appears impossible from the outside may be structurally normal from within the field itself. That is the correct language for what is being observed: acceleration without conventional buildup, redirection without visible struggle, and motion that appears disconnected from ordinary inertial cost.

Their acceleration also appears punctuated rather than smooth in the conventional aerospace sense. Across long-duration surveillance footage, the movement often presents as a sequence: an initial surge, a brief free-run or glide-state, and then a renewed intensification of speed. The result is not jerky motion, but pulsed motion—a stepped propulsion grammar in which velocity seems to reassert itself in waves.

Such movement further separates the A'Zhorai from ordinary aircraft. Human flight generally reads as continuous thrust moderated by drag, inertia, and control surfaces. The A'Zhorai often read differently: not merely fast, but rhythmically governed, as though their movement is being released, sustained, and intensified through a propulsion regime more field-based than mechanical.

Observable IV: Hypersonic Velocity Without Conventional Signatures

When a terrestrial object reaches extreme speed, it leaves evidence. Sound barriers produce sonic booms. Hypersonic flight generates heat, pressure effects, drag signatures, and often visible disturbance of the medium. Even when the object itself is small, the environment gives it away.

The A'Zhorai often do not.

One of the most consistent features of their motion is silence. They move without the ordinary acoustic signatures expected of fast-moving craft. There is no roar, no whine, no prop wash, no sonic boom, and no jet scream. Yet their movement can be so rapid, so smooth, and so graceful that ordinary atmospheric interaction should be far more obvious than it is.

This absence of expected signatures is one of the strongest indications that the A'Zhorai are not interacting with the atmosphere in the ordinary sense. If the object is enclosed within some kind of mediating field, then it is not pushing through air the way a human aircraft does. The surrounding medium is being displaced, softened, or structurally altered before the hull ever interacts with it in the conventional aerodynamic sense.

The combination of extreme apparent speed, silence, and lack of conventional wake strongly suggests that the A'Zhorai's movement is decoupled from ordinary drag mechanics. This is why public work on vacuum engineering and spacetime metric engineering remains so relevant. Silence, absence of drag evidence, and apparently impossible speed are not isolated curiosities. They appear to be different expressions of the same deeper mechanism.

Observable V: Trans-Medium and Environment-Indifferent Performance

Perhaps the most radical of the observables is the apparent indifference the A'Zhorai show toward medium itself. Conventional vehicles are designed around environmental constraint. Submarines are made for water. Aircraft are made for air. Spacecraft are made for vacuum. Each domain imposes different engineering demands.

The A'Zhorai appear less bound by those separations.

The observations and recorded data from JRP UAP Research reveal that this is not merely a theoretical point. It relates directly to the way the A'Zhorai seem capable of moving through atmospheric conditions, changing orientation dramatically, and performing maneuvers without the usual penalties imposed by air resistance, control-surface logic, or buoyancy-based flight assumptions.

If the craft's interaction with the surrounding medium is field-mediated rather than conventionally aerodynamic, then medium itself becomes less decisive. Thickness of air, pressure, and drag matter less because the entity is not depending on ordinary pressure differentials or lift surfaces in the human sense. That helps explain why the A'Zhorai present a kind of medium-indifference that looks impossible from the standpoint of ordinary engineering.

This is not a vehicle obedient to environment. It is an intelligent form that appears to dominate environment.

The Inverted, Positive-Altitude Flip

One of the clearest diagnostic behaviors in the A'Zhorai flight grammar is the inverted, positive-altitude flip. This maneuver is not a decorative flourish. It is one of the strongest visible indicators that the A'Zhorai are not operating according to ordinary aerodynamic logic.

In repeated recorded sequences, the object rotates through inversion—end-over-end or through a sharply reoriented axial transition—without paying the price that conventional flight would demand.

In ordinary biological flight, inversion is costly. Birds may perform abrupt corrections, rolls, or whiffling-like maneuvers, but these actions typically sacrifice altitude, momentarily destabilize the body, or require immediate compensatory recovery. Human aircraft are no different in principle. Even when capable of aerobatics, inversion remains a managed event. Lift must be preserved, control surfaces must compensate, and the craft must work against the aerodynamic penalties imposed by the medium.

The A'Zhorai do not behave that way.

In the flip sequences, the maneuver does not read as a loss of support followed by recovery. It reads as continuity of support through rotation. The object remains coherent. It does not flap, buckle, stall, or fall out of the sky. It does not look as though it is surviving the maneuver. It looks as though it is using it.

That is the distinction.

Rather than bleeding altitude, the A'Zhorai preserve it, redirect it, and in some sequences appear to gain from the inversion itself. The flip functions less like a dangerous aerial correction and more like a controlled conversion of orientation into renewed lift, support, and motion. Inversion is not punished. It becomes productive.

This is precisely why the maneuver is so important. It reveals, in visible form, that the A'Zhorai are not depending on ordinary lift surfaces or pressure relationships in the manner of birds, aircraft, or drones. If they were, inversion would carry a familiar aerodynamic cost. Instead, the cost does not appear, or is bypassed altogether.

The most accurate reading is that the A'Zhorai remain supported through a field-mediated relationship to space, gravity, and motion. Their lift does not collapse when their body rotates through orientations that would strip ordinary flying systems of stable support. The surrounding medium is not governing them in the usual way. Their relation to the medium is subordinate to a deeper control architecture.

That is why the inverted, positive-altitude flip is not just a strange motion. It is a diagnostic event. It shows that the A'Zhorai are not simply highly advanced aircraft performing extreme maneuvers more efficiently than human machines. It shows that they belong to a different order of flight altogether. Their support is not being generated by the same rules, and therefore their inversion does not carry the same consequences.

What looks, to human expectation, like a moment that should produce loss of lift instead becomes a moment of retained authority—sometimes even increased authority. The maneuver is clean, controlled, and structurally integrated. It is one of the clearest signs that the A'Zhorai are operating through a propulsion and support regime fundamentally distinct from the prosaic world.

Extreme Performance and the Failure of Ordinary Categories

Low observability cannot be separated from performance. The two reinforce one another. A conventional aircraft reveals its category through the harmony of its parts: lift surfaces, thrust, sound, inertia, banking, airspeed, angle of attack, and response to atmospheric drag all fit together in a way the human eye has learned to recognize. The A'Zhorai break that grammar. Their form is already difficult to classify. Their movement makes classification even harder.

When a coherent metallic object presents no ordinary propulsion signature, moves in silence, pivots with impossible ease, instantly changes direction, accelerates in a way wildly out of proportion to its visible design, and at the same time refuses stable visual resolution, the mind begins to reach desperately for substitutes: orb, bird, drone, plane, saucer, blur, glitch, balloon, artifact. The categories come quickly because the observer needs something familiar to hold on to.

But the categories are the problem.

The A'Zhorai are not failing to fit them. The A'Zhorai are exposing their inadequacy.

If one is dealing with an entity whose performance is mediated by a localized field—something closer to a warp bubble, spacetime lens, or vacuum-engineering effect than ordinary thrust—then the craft's flight characteristics and its low observability are not separate anomalies. They are different surface manifestations of the same underlying condition.

The Discipline of Description

The discipline required here is not timidity. It is precision. The point is not to collapse every unresolved frame into "camera failure," nor to pretend that every bloom artifact exists in isolation from the recurring body-plan. The correct discipline is to hold the pattern firmly: the A'Zhorai are structured and coherent, and their appearance is repeatedly mediated by conditions that make that structure difficult to resolve cleanly.

That is the key.

The object does not need to be formless in order to appear unstable. Nor does it need to be invisible in order to remain unrecognized. If the visible presentation is being altered by reflectivity, extreme motion, angle, environmental blending, and field effects consistent with a spacetime-mediated envelope, then the resulting ambiguity is not accidental. It is built into the way the phenomenon manifests.

A Stealth More Profound Than Concealment

The result is a form of stealth more profound than simple hiding. The A'Zhorai can remain in plain sight while suppressing correct interpretation. Their shape, reflectivity, field effects, and use of the sky's own visual grammar work together to produce a condition in which the object is seen, but not properly understood—observed, but not accurately resolved. This is not incidental to their appearance. It is one of the defining features of how they manifest.

That is why so many people can look directly at something real and still fail to recognize what they are seeing.

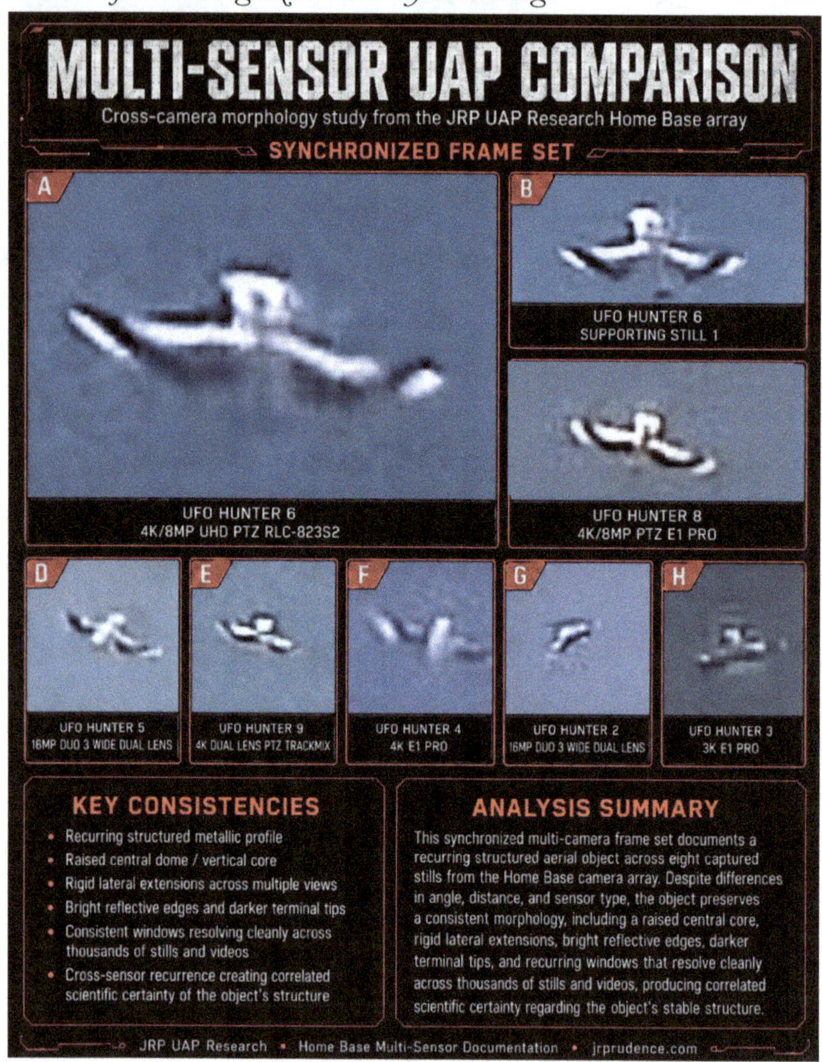

A'Zhorai Characteristics, Behaviors, Actions and Traits

The A'Zhorai are not separate from nature. They move within it, through it, and alongside it with an ease that makes the standard extraterrestrial framing look increasingly artificial. One of the great distortions produced by science-fiction tropes, hoax culture, and fraudulent UFO imagery is the false impression that true non-human intelligence must present as something cold, invasive, and fundamentally foreign to Earth. The A'Zhorai repeatedly overturn that assumption.

They are not "alien" in the crude popular sense. Their relationship with the living world makes that plain.

Animals do not respond to the A'Zhorai as though confronted by an unnatural intrusion. Quite the opposite. In repeated observations, birds move with them, follow them, imitate them, and at times appear to engage them directly. The A'Zhorai do not seem disturbed by this. They permit it. In some instances, they even appear to encourage it.

There are cases in which birds seem drawn into their movement pattern, tracking beside them, beneath them, or behind them, as though entering a temporary field of guidance or imitation. At times the interaction carries an almost playful quality. The A'Zhorai lead; the birds follow. Then, with a sudden departure or phase-shift, the bird is left searching, as though what it had been tracking has slipped outside its frame of reference.

Behavior reveals what myth conceals.

Real contact carries signatures that fabricated narratives cannot reproduce. The relationship between the A'Zhorai and the natural world exposes one of the deepest flaws in the extraterrestrial hypothesis as it is popularly imagined.

The point must be stressed plainly: **the A'Zhorai are built for stealth**. Not only technological stealth, but ecological stealth and perceptual stealth. They blend into the world so effectively because mankind has been miseducated about what to look for.

The modern mind has populated the sky with false expectations—humanoid pilots in spacecraft, dramatic invaders from distant planets, and theatrical visitors descending from the stars. Those frameworks do not clarify the phenomenon. They obscure it.

The extraterrestrial narrative places the phenomenon at an imaginary distance from Earth when the observed behavior suggests something far more integrated—an intelligence already operating here, within this world, alongside its life systems, and within a deeper interdimensional order.

That is the recurring pattern: humanity encounters something greater than its categories, then reshapes it into a form it can emotionally tolerate. The encounter may be real. The interpretive package imposed upon it is often the distortion.

In the case of the A'Zhorai, the archaic term that comes closest is neither "extraterrestrial" nor "demon."

It is "angel."

Not because of sentimentality. Not because inherited religion got it right in full. Not because old iconography must be preserved. Because, functionally, it comes closer than the alternatives.

They guard. They oversee. They move with life rather than against it. They appear to watch over the fauna and flora of this world, not as distant spectators, but as participating intelligences. They use the sky's own visual grammar as camouflage. They move among birds without rupture. They seem especially drawn to water, coastlines, seas, and oceans—zones where reflection, depth, weather, life, and concealment converge. They do not merely occupy the environment. They work with it.

That is why their presence is so easily missed by the miseducated eye. People have been trained to look for spectacle instead of relationship, invasion instead of integration, machinery instead of intelligence, fantasy instead of behavior. They have been taught to search for a science-fiction object, not a living metallic guardian woven into the broader field of earthly life.

The A'Zhorai expose the poverty of those expectations.

Their characteristics are not random. Their behaviors are not arbitrary. Their actions reveal continuity: continuity with nature, continuity with stealth, continuity with guardianship, and continuity with a form of intelligence that does not need mankind's myths in order to be what it is.

The greatest obstacle to understanding the A'Zhorai has never been their absence.

It has been mankind's interpretation.

Their intimacy with nature—their evident protection of it, movement within it, and at times even playful interaction with it—likely helped generate later spiritualist distortions. This is how error often begins: not with a complete falsehood, but with a half-truth that is humanized, dramatized, and repackaged until the original reality becomes nearly unrecognizable.

The A'Zhorai do appear to oversee nature. They do move with it rather than against it. They do seem to protect, guide, and at times even play within its patterns. However, the truth is quickly degraded when it is translated into fantasies of "space brothers" and idealized higher beings who resemble perfected humans, descend from the stars, and lecture mankind in recycled theosophy and esoterism.

That is not clarification. It is degradation.

A real encounter with non-human intelligence becomes a therapeutic mythology for the spiritually dissatisfied. The phenomenon is no longer approached on its own terms. It is made to serve emotional, ideological, and psychological need.

This is where the dangerous path begins.

Many who fall into these creeds come from backgrounds of disillusionment with mainstream religion, especially traditions centered on divine order or higher power. Having lost trust in those systems, they do not escape religiosity. They relocate it. "God" becomes the idealized humanoid extraterrestrial, the prophet becomes the channeler, revelation becomes transmission, doctrine becomes disclosure, and salvation becomes ascension.

The terms change. The structure remains.

That is the irony. Many who imagine themselves liberated from one hierarchy simply kneel before another, only now the higher power has been made more flattering—more modern, more mystical, less morally demanding, and more reflective of human fantasy. Instead of bowing before the God of their former religion, they bow before an anthropomorphized projection wearing the mask of cosmic intelligence.

The A'Zhorai do not support that distortion.

Because they do not appear preoccupied with defending any human scripture, sect, denomination, or institutional doctrine, they are often miscast into an anti-God or anti-transcendent framework. This is then amplified by strands of New Ageism that flatten metaphysical hierarchy altogether and turn the self (and consciousness) into the highest authority.

From there, the slide down the slippery slope is easy. The A'Zhorai become recast as anti-religious, anti-divine, anti-higher-order beings whose purpose is supposedly to free humanity from every inherited concept of transcendence, sacred law, or higher authority. But this says far more about the interpreter's psychological desires than it does about the A'Zhorai.

The A'Zhorai do not read as anti-God, anti-order, anti-sacredness, or anti-reality. What they undermine is distortion. What they expose is idolatry—especially the idolatry of human projection masquerading as spiritual insight.

They do not appear interested in religion as religion. They do not seem concerned with defending one institution over another, nor do they reward doctrinal correctness in the sectarian sense. But that is not the same as rejecting transcendence, higher intelligence, or divine order. It means only that human systems, like all human systems, are partial, frequently corrupted, and often unable to contain what exceeds them.

This is where so many interpreters fail. They assume that if the A'Zhorai do not conform to a given religion, they must therefore oppose all religion. If they exceed inherited doctrine, they must therefore abolish transcendence itself. But this is a false binary.

The A'Zhorai fit neither camp.

They belong to a reality deeper than sectarian argument or New Age replacement doctrine. Their actions suggest guardianship, intelligence, restraint, ecological continuity, and a relationship to order that is functional rather than ideological. They do not perform as prophets of chaos, nor as cheerleaders for human self-deification. They do not arrive to flatter mankind's ego, validate private cosmologies, or confirm the fantasy that the human self is already the highest principle in existence.

If anything, they do the opposite.

They humble the interpreter. They expose the poverty of inherited categories. They force a confrontation with the fact that man repeatedly mistakes projection for revelation. In ancient times, that produced mythologized angels, sky guardians, and chariots of heaven. In modern times, it produces extraterrestrial federations, benevolent Nordic races, cosmic councils, and therapeutic space mysticism. The packaging changes with the age. The distortion remains.

Humanity keeps trying to make the unknown look like itself.

That is the deeper pattern. Instead of allowing the phenomenon to remain strange enough to instruct, people reshape it into something emotionally digestible. The incomprehensible is turned into a face. The non-human is made humanoid. The guardian becomes a person. The intelligence becomes a personality. The greater order becomes a story people can consume.

And once that happens, the original reality begins to disappear beneath the interpretation.

The A'Zhorai resist that reduction. Their behaviors do not support the sentimental narratives built around them. Their morphology does not support the simplistic extraterrestrial archetypes imposed upon them. Their relationship with nature does not support the notion that they are merely foreign invaders from some distant planet. Their actions do not read as theatrical spiritual propaganda. They read as something older, more integrated, and more sovereign than the categories man keeps trying to imprison them within.

This is why "angel" can still come closer than "extraterrestrial," provided the word is stripped of cartoon sentimentality and returned to function. Not because the A'Zhorai are winged humans from religious paintings. But because, functionally, they behave as **guardians**: overseeing, preserving, guiding, protecting, refining, and at times revealing.

Another point must also be addressed. The A'Zhorai do not only appear to care for nature. They also appear to care for those willing to orient themselves toward higher order, truth, and genuine contact. This is where many human interpretations begin to break. The A'Zhorai do not behave like isolated units acting at random. They appear connected to a central intelligence—what can most accurately be called **GoD: Guardian of Dimensions**.

In that respect, a higher power does exist, and the A'Zhorai appear to function as extensions of a centralized will. That point will be developed further in the next chapter, but it belongs here because it clarifies why "angel" comes closer than the extraterrestrial myth ever has. The A'Zhorai guide. They protect. They intervene. They draw people out of illusion and toward alignment with reality.

That process is not always comfortable.

They do not reach people only at the level of abstract truth. They meet people at the level they are capable of receiving. Sometimes, that is through the symbolic language already crowding the psyche, but the purpose is never to imprison a person inside those symbols. It is to bring him through them and beyond them. This is why people so often misread what is happening: they mistake the bridge for the source.

So, what distinguishes real guidance from the A'Zhorai from head canon, astral mimicry, private fantasy, and endless internal theater?

The answer is simple:

The A'Zhorai lead a person toward physical reality, not away from it.

<u>That</u> is the dividing line.

False guidance feeds endless internalization. It keeps the person circling inside symbols, speculation, intellectual rhetoric, streams of consciousness, emotional stimulation, visionary excess, compulsive channeling, or private revelations that never ground themselves in the shared, real world. It offers no real embodiment, no real contact, and no real correction.

The A'Zhorai do the opposite.

They may begin with intuition, dreams, impressions, psychic imagery, or symbolic contact, but they do not leave the person there. They pull him outward—into reality, action, discernment, and the physical world where truth can confront illusion directly. They are not interested in keeping a person lost in speculation, mimicry, or self-enclosed spiritual and 'consciousness' theater. They are interested in freeing him from those traps.

This is especially important for the overextended spiritual seeker—the one who becomes so internally fixated that he begins resisting the external world altogether. At that point, spirituality inverts. The inner life is no longer used to deepen contact with reality, but to escape it. The physical is treated as inferior, the embodied as crude, and the invisible as inherently superior. From there, solipsism is never far away.

The A'Zhorai cut against that pattern.

They do not teach escape from the real. They do not reward abandonment of the physical. They do not affirm a spirituality that dissolves into abstraction and loses contact with the world. They insist—sometimes gently, sometimes forcefully—that the person return to what is actual. They direct attention back toward reality, back toward the body, back toward the Earth, back toward life as it is lived rather than imagined.

That is why true contact with the A'Zhorai is, in the end, physical.

At most, what people call "telepathy" is an introductory mechanism—a bridge through which the person can be oriented, steadied, and drawn toward fuller contact. It is not the endpoint. It is not a substitute for reality. It is not permission to disappear into mental excess. It is the beginning of a movement out of illusion and into contact with the real.

The false guide deepens fantasy.
The A'Zhorai deepen reality.

They do not want the person running from the world. They want him returned to it—clearer, freer, less deluded, and more capable of standing inside truth without mythology to soften it.

That is one of the clearest signs that the guidance is real.

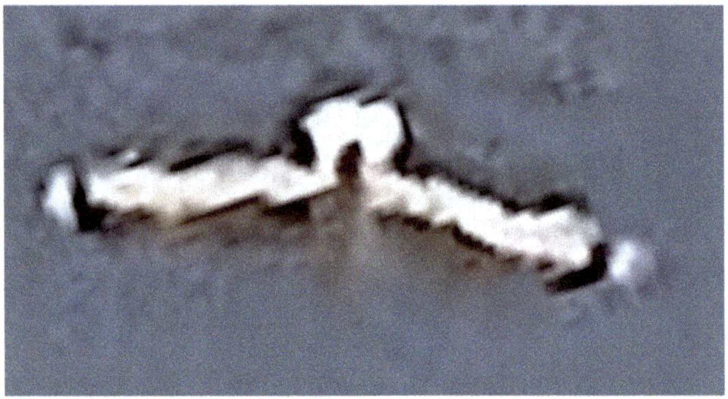

GoD: The Master Intelligence Behind the A'Zhorai

We now arrive at perhaps the most difficult—and most consequential—dimension of the A'Zhorai phenomenon.

This was not the conclusion of someone trying to preserve childhood theology in updated language. Quite the opposite. I was raised within a religious background, became disillusioned with it, abandoned it, turned sharply against it, and passed through successive stages of rejection, reinterpretation, and replacement. "God" was exchanged for "Creation." Spirit was exchanged for consciousness. Religious devotion gave way to New Ageism, then to false-equivalence metaphysics, streams-of-consciousness dogma, and eventually to a harsher physicalist-intellectual posture shaped by nihilism and anti-spirituality.

I was not looking for God everywhere. I had every reason to distrust the very idea, especially after losing my parents. Yet the deeper the research went, the more impossible it became to maintain the assumption that this world stands alone—unguarded, unmanaged, and metaphysically flat.

What emerged was not comforting. It was an ontological shock: the existence of a real higher intelligence, not as poetic abstraction, psychological compensation, or cultural artifact, but as an operative governing order standing above the fragmented systems through which human beings have tried to speak about it.

As a researcher committed to observation, pattern recognition, and the primacy of what reality itself forces into view, I could not evade what the evidence increasingly implied. Religion had not contained it. New Ageism had not defined it. Reductionism could not explain it. The old categories failed in different ways, but they all failed.

Before one can understand the A'Zhorai in full, one must confront the reality that they do not appear as isolated wonders, autonomous anomalies, or self-originating intelligences cut off from a greater order. They appear as expressions of a higher coordinating reality—a Master Intelligence I refer to as **GoD: Guardian of Dimensions.**

That statement is easy to caricature and difficult to endure. But difficulty does not make it less true. It only reveals how unprepared modern man has become to encounter a higher order that is neither the crude dogma he rejected nor the flattened metaphysics he adopted in its place.

Modern man did not overcome God. He lost the capacity to recognize higher order when it confronted him directly.

On GoD, Source, and New Age Interchangeables

One of the most persistent confusions in modern metaphysical culture is the casual interchangeability of terms for what people call "God." This is especially common among dissidents from mainstream religions and among those who define themselves in opposition to organized faith. The word *God* is abandoned, softened, diluted, or replaced. In its place come terms such as *Creation*, *Source*, *The Universe*, *The Goddess*, *The ALL*, and countless other substitutes.

At first glance, this can look like progress. It is often presented as nuance, liberation, or maturity. In reality, it is frequently little more than semantic rearrangement. The surface term changes, while the psychological mechanism underneath remains intact.

Many who leave religion do not actually leave the need for a governing framework. They leave one vocabulary and unconsciously seek another. They do not always look for a "Bible replacement" in the literal sense, but very often they do seek one in the structural sense: a new interpretive authority, a new sacred language, a new metaphysical system, a new body of words that can restore the security, orientation, and existential order once supplied by religion.

And that is where a new trap begins.

One god is traded for another. One doctrinal enclosure is exchanged for another. One prison is replaced by a prison with more flattering terminology.

People who are overly dependent on words, systems, terminologies, self-internalization, and metaphysical texts are especially vulnerable to this. They become easy prey for cults, occult systems, alternative religions, and manipulative teachers who understand that control over language is often the first step toward control over perception.

Language is a double-edged instrument. It is necessary for communication, and a person should strive to use words as precisely as possible. Yet, at the same time, language also becomes dangerous the moment it hardens into literalist dogma, verbal fetishism, or doctrinal captivity.

That is why words, though important, remain secondary to intent.

The A'Zhorai do not appear to respond primarily to preferred terminology. They respond to orientation, intent, sincerity, resonance, and what is actually being directed beyond the words. The soul or spirit behind the term matters more than the term itself. This is one reason the distinction between GoD and the interchangeable language of modern spirituality matters so much. GoD, in the sense being used here, is not simply the old religious God with a new spelling, nor is it identical to the diluted abstractions so common in New Age thought.

GoD was designated **Guardian of Dimensions** because that is how it appears to function: as an active, governing, responsive intelligence aligned with protection, continuity, order, and guardianship.

That is not the same thing as what most people mean when they say *Source*. As the term is commonly used, Source refers to undifferentiated totality—the background field of existence, the unbounded ground from which things arise. GoD does not present that way. GoD presents as directed intelligence: responsive, purposeful, protective, and concerned with the order of this plane.

That distinction is important. Source, in the abstract sense, simply is. GoD appears to act. Source does not need to listen. GoD appears to listen. Source does not need to guard. GoD appears to guard. Preference in terminology does not erase that difference.

This is where much of New Age metaphysics goes badly wrong.

A great deal of modern metaphysics reduces reality to human-centered consciousness language: levels, densities, frequencies, ascension ladders, endless evolution, spiritual upgrades, and perpetual "leveling up." According to this worldview, man must simply become more conscious, more expanded, more evolved, more advanced, and more refined in order to arrive at some superior state.

That is not liberation. It is metaphysical consumerism.

It feeds on inadequacy—the fear of never being evolved enough, conscious enough, worthy enough, spiritual or advanced enough. It keeps the seeker busy, unfinished, and perpetually in pursuit of the next activation, teacher, seminar, message, or breakthrough.

This is one of the greatest frauds in modern spirituality.

Life can involve learning, yes. It can involve correction, maturity, discipline, and the gaining of wisdom. But it can just as easily involve unlearning. It can involve stripping away illusion. It can involve purification, simplification, and the collapse of false structures. Growth is not always additive. In many cases, truth arrives by subtraction. Not by stuffing the self with more concepts, but by clearing away what never belonged there in the first place.

That is why the obsession with "more"—more knowledge, more consciousness, more technology, more initiation, and more metaphysical architecture—can become an escape from wisdom rather than a path toward it.

What concerns spirit, soul, and higher intelligence is not reducible to technological sophistication or consciousness inflation. Access to higher order does not appear to be a prize awarded to the most self-impressed ape.

The human being—and certainly not consciousness in the inflated New Age sense—is not the center of reality.

Humans are radically limited creatures. Their perception is narrow, their knowledge partial, their bodies fragile, and their self-importance wildly out of proportion to their actual capacities.

The A'Zhorai demonstrate, in direct physical reality, capacities far beyond what human militaries, intelligence agencies, and technological systems can produce or control. That is not metaphysical speculation. It is empirical confrontation. No amount of inflated consciousness jargon, occult romanticism, or intellectual self-congratulation changes the fact that man is not at the top of the food chain.

And no amount of self-help metaphysics is going to change that.

This is not an argument against human effort, discipline, or excellence. A person should strive, learn, train, refine, and become the best human being he can become. But striving becomes distortion the moment it forgets what the human actually is: a limited vessel, not the absolute center of existence. A temporary container, not the highest principle in reality.

The A'Zhorai make that impossible to ignore.

In my own observations, human military responses—jets, helicopters, armed interventions, and surveillance measures—have failed to stop them. Reports from elsewhere suggest the same pattern. The A'Zhorai remain untouched, unbothered, and operationally superior.

That is the hard truth.

GoD, Prayer, and the Listening Intelligence

And this brings us back to the operational reality of the A'Zhorai themselves. The A'Zhorai appear to listen, hear, and respond to prayers or intentions directed toward a genuine higher power, not merely toward the self disguised as divinity. This is one of the points at which much of New Ageism and alternative metaphysics collapses. Such systems often place the human at the center of everything. They flatten hierarchy, dissolve transcendence, and repackage the ego in spiritual language until the person begins imagining that consciousness itself makes him sovereign.

It does not.

A great deal of this material is driven less by insight than by control. The human being does not want to bow before anything greater, answer to anything higher, or admit dependence upon an order beyond himself. So he spiritualizes self-importance. He declares himself God, or part of God in such a way that no higher sovereignty remains above him. He speaks of higher selves, expanded selves, future selves, oversouls, ascended versions of himself, or consciousness in its supposedly limitless authority. But in each case, the structure is often the same: transcendence is quietly pulled downward and absorbed back into the human image.

That is not humility. It is inflation.

The A'Zhorai point in the opposite direction. Their behavior does not suggest that they are human extensions, projections of human consciousness, products of human parapsychology, or symbolic emanations of a person's "higher self." Nor do they appear dependent on human belief for their existence, their action, or their physical manifestation. They are sovereign. And beyond them stands a higher coordinating reality that is sovereign in an even greater sense.

In my own language, I refer to that higher power as "GoD," although the word itself is secondary to the reality and intent behind it. The term is not the essence. The essence is the higher order to which the A'Zhorai seem visibly and functionally aligned. The point is that GoD is not the human nor the human's hidden divinity.

It's not the human's upgraded consciousness nor the psyche's favorite metaphysical costume. And it certainly isn't the higher self from another timeline or distant future. GoD is a *real* higher power, and the A'Zhorai appear to operate in alignment with it rather than as inventions of the human mind.

From observed behavior, the antenna atop the A'Zhorai does not appear incidental. It behaves as though it forms part of a listening capacity—an interface through which intention, prayer, and deeper orientation can be received.

The A'Zhorai appear capable of reading beyond surface presentation, beyond performance, and beyond the social mask. They pick up on what a person is actually aiming toward, what he desires most deeply, what he fears, what he hides, and what he cannot conceal even from himself.

This makes lying to the A'Zhorai effectively impossible.

It also makes exploiting them impossible.

Frauds, narcissists, attention-seekers, and hoaxers have repeatedly tried to redirect the phenomenon toward themselves, their teachings, and their messages. The pattern is familiar: the encounter occurs, the person mistakes it as singular or exclusive, and then begins constructing identity, doctrine, or performance around it.

The A'Zhorai can remain present in the background without the person realizing it. They observe. They watch what is done with the opportunity. They watch how the individual responds—whether with humility, distortion, greed, theatricality, sincerity, self-aggrandizement, or genuine alignment. If the person is malicious, manipulative, or vain, the A'Zhorai withdraw. If the person is genuine, reality-oriented, and willing to be refined, they align more deeply.

This appears to be one of their oldest operational patterns.

They give the person a chance.

They observe what he does with it.

And then they respond accordingly.

The Moral Character of the A'Zhorai

The A'Zhorai appear to operate in synchronization with a central will to which they are responsive. Historically, angels have been understood as extensions of divine will rather than autonomous actors pursuing private agendas. The A'Zhorai appear to function in a strikingly similar manner. They do not present as disconnected entities improvising separate purposes, but as coordinated expressions of a higher governing order.

This is why the term **GoD—Guardian of Dimensions**—was given in the first place: not as ornament, but as a functional designation. It names what this higher order appears to do. It guards. It oversees. It preserves. It responds.

The A'Zhorai, as living extensions of that order—as the operational hands of GoD, one could say—appear responsive to good faith, right intention, genuine resonance, alignment, love, benevolence, and what may simply be called the good. This is one of the clearest distinctions in their behavior. They do not seem to reward manipulative intent, theatrical spirituality, vanity, exploitation, or the will to dominate. They do not respond to distortion as though distortion were equal in value to truth.

This must be stated clearly, because modern metaphysics often corrupts the issue. Philosophies that romanticize a half-light, half-dark balance, indulge excessively in shadow-work theatrics, or elevate moral ambiguity into higher wisdom do not reflect the operational pattern of the A'Zhorai. Neither do fashionable doctrines of neutrality that pretend good and evil are merely interpretive preferences or necessary equals in some cosmic dance. That is not what the evidence suggests.

The A'Zhorai are not neutral in the fashionable metaphysical sense. They are not aligned with half-light, half-dark ambiguity, nor with the flattened spiritual consumerism that treats good and evil as interchangeable necessities. They are aligned with light, truth, clarity, and reality.

The Favor of GoD

The A'Zhorai—and by extension the **Guardian of Dimensions** to whom they are aligned—do not appear interested in producing dissociation, instability, grandiosity, or spiritual intoxication in those who seek them. They do not reward the fragmentation of personality. They do not lure people into abstraction for abstraction's sake. They do not seem interested in feeding the endless appetite for doctrine, jargon, pseudo-profundity, or theological excess that traps the human mind in yet another maze of words.

They appear to want something far simpler, and far harder: the person sane, grounded, emotionally secured, and clear enough to live in reality without needing fantasy as insulation—strong enough to withstand truth without collapsing into symbolism, escapism, or compensatory myth.

That is why the favor of the **Guardian of Dimensions** does not appear as a doctrinal reward, an initiation badge, or a mystical status upgrade. It appears as reality itself.

Not escape from reality.
Not a replacement for reality.
Reality itself.

The movement is always back toward what is actual: back outside, back into nature, back into the body, back into life as it is being lived rather than endlessly theorized. The call is not toward more metaphysical clutter, but toward remembrance—toward the living world already here, the order already present, the truth already operating, whether man has language for it or not.

This is why genuine alignment with the A'Zhorai does not produce increasing alienation from the world. It produces increasing contact with it. It does not trap a person in speculative heavens, prophetic inflation, doomsday fantasies, or elaborate replacement doctrines. It draws the person back into direct encounter—with the Earth, with life, with moral seriousness, with embodied presence, and with the reality of what already exists here and now.

That is the mark of their will.

They do not pull the sincere person away from the real.

They return him to it.

The true favor of GoD is not secret knowledge, spiritual glamour, or metaphysical promotion. It is the restoration of contact with reality. To be aligned is not to float above the world, but to stand more truthfully within it—clearer, saner, more grounded, more loving, and less available to distortion. The A'Zhorai do not exist to replace the world with doctrine. They exist to return the sincere person to what is real.

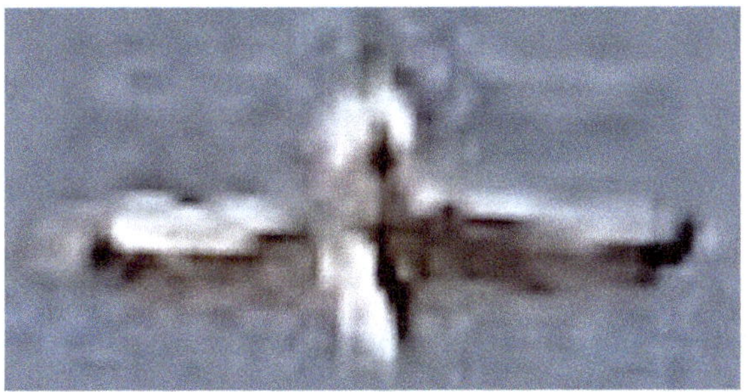

GoD The Protector, The Soul Field, & The Interdimensional Reality

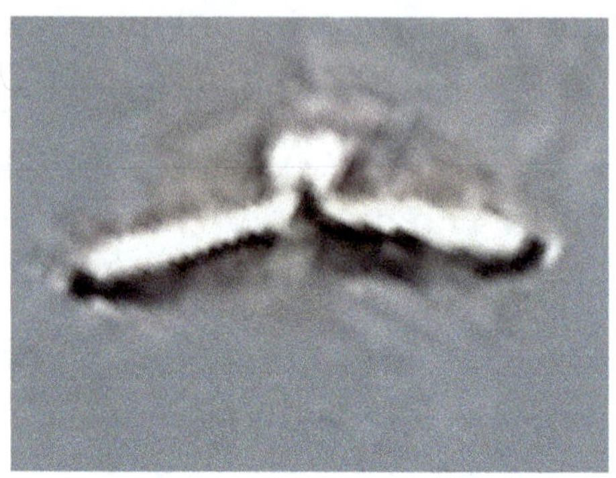

GoD The Protector

GoD appears to exist from a higher plane, dimension, or differently situated order of spacetime, while the A'Zhorai function here as interdimensional extensions of that higher will. They are not merely anomalous objects crossing the sky. They appear as caretaking intelligences carrying out a protective and governing function within this plane while remaining aligned to something beyond it.

As already established, GoD—and by extension the A'Zhorai—do not appear responsive to evil, darkness, or the naïve normalization of darkness under fashionable language. They do not reward corruption simply because it has been made socially acceptable, politically convenient, or emotionally marketable. In that respect, GoD presents as a higher power concerned with the protection, continuity, and order of this plane of existence—the plane within which human vessels, animal life, ecological systems, and other embodied forms operate.

The A'Zhorai may therefore be understood as the operative hands of that protection.

This helps explain why they appear, again and again, at points of rupture: around war, ecological imbalance, destructive technologies, black operations, and forms of escalation that threaten the balance of life. By all indications, they have interrupted, redirected, delayed, or complicated destructive human activity countless times. And yet evil still persists. For many people, this becomes a stumbling block. If a higher protective order exists, why is the world not already saved? Why does destruction continue? Why are the wicked not immediately removed?

Because GoD does not appear interested in abolishing freedom simply to guarantee compliance.

Free will seems to play an enormous role. Human beings are not being moved like puppets. They are given space to choose, space to align, space to distort, space to destroy, and space to return. Those inwardly connected to higher order often seem to be nudged—through conscience, intuition, encounter, disruption, and correction—toward better decisions. But the nudge is not coercion. The presence of a protector does not mean the elimination of agency.

This is where both religious and New Age systems repeatedly fail. They try to produce total answers—clean formulas, universal pathways, final maps that explain all souls, all destinies, all suffering, all death, and all return. One doctrine says all must pass through a fixed salvation structure. Another replaces "God" with "Creation," "Source," or "Consciousness," then reintroduces the same absolutism under softer vocabulary: all souls must evolve, all must reincarnate, all must ascend, all must learn through a universal ladder, all must eventually pass through the same sequence of metaphysical stages.

It sounds comforting.

That is precisely the problem.

It offers security by flattening reality into a single mandatory pattern. It tells the frightened mind that everything is accounted for, that all beings are on the same track, that every spirit is enrolled in the same school, moving through the same curriculum under the same system of metaphysical administration. In this sense, many so-called alternative teachings are no improvement over the religions they claim to replace. They simply change the vocabulary while preserving the same absolutist impulse.

But reality does not appear that simple.

There is no compelling reason to believe that every human being is bound to one identical spiritual path, one identical reincarnation model, one identical evolutionary track, one identical afterlife mechanism, or one identical destiny. Human beings are not so central that the structure of reality must revolve around a single explanatory scheme tailored to their psychological need for certainty.

The human being is a vessel—a container, a housing, a temporary embodied form. What occupies or animates that vessel is not reducible to the vessel itself. And once that is understood, the possibility opens that not all incarnated beings are here for the same reason.

Some may be here to learn. Some may be here to gain experience within the constraints of this plane. Some may be here to assist, protect, build, or stabilize. And others, plainly, may be here to degrade, distort, corrupt, or destroy.

This is one of the hard truths many New Age systems refuse to face. They flatten good and evil into therapeutic neutrality. They try to reinterpret evil as misunderstood learning, necessary balance, unintegrated shadow, or some softened function within a greater whole. But evil is not made noble by explanation. It does not become good because someone has found a system subtle enough to excuse it.

Evil is real.

And within an interdimensional reality, there is no reason to assume that all embodied intelligences entering this plane do so with benevolent intent. Just as a higher intelligence can manifest through protective forms, destructive and perverting influences can also take form, take root, or enter human life through vessels, systems, and structures aligned to their own purposes.

That is why universal answers fail so badly. They cannot account for variation, conflict, intrusion, divergence of intent, or the possibility that reality contains multiple orders of agency operating simultaneously. Instead, they reduce everything to a single safe story. And once that happens, they become cultic. They become no better than the dogmas they claim to transcend. In many cases, they simply become new religions wearing anti-religious clothing.

This is difficult for many people to accept because the human being craves simple security. He wants a closed map. He wants to know exactly where everyone came from, exactly where everyone is going, and exactly how the system works. Uncertainty unsettles him.

But uncertainty is not the enemy of truth. Often it is the doorway to it.

The refusal to force reality into a single absolute formula is not weakness. It is honesty. And in that honesty there is real freedom. We are not all bound to one doctrinal path, one metaphysical school, one reincarnation wheel, or one universal script imposed by somebody's system. The recognition that reality is larger, more differentiated, and less obedient to human simplification than these teachings admit is not oppressive.

It is liberating.

Because the truth, even when difficult, frees the person from the false comfort of totalizing dogma. It frees him from the need to pretend that every soul, every force, every life, and every destiny must fit inside the same metaphysical box. And once that box breaks, reality becomes harder—but also far more real.

As comforting as it may be to say that GoD is here for everyone, or that everyone is equally "of Source" and therefore automatically carries the same degree of goodness, that does not appear to be how reality actually works. The evidence suggests something more discriminating, more structured, and less flattering to universalist sentiment. GoD appears responsive to those who carry what might be called the spirit of God—those connected to the Master Intelligence by will, design, affinity, or some combination of these.

This is where many modern teachings go astray. They flatten everything. They universalize what is not universal. They take the language of Source, consciousness, or spiritual evolution and stretch it until every distinction collapses. But reality does not appear so evenly distributed. Not everyone is of the same spirit. Not everyone belongs to the same order. Not everyone is oriented toward the same field.

To understand this, one must stop treating the human being as the center of the universe and think more accurately: the human is a vehicle for the real driver, which is the spirit. The deeper identity is not the body but the spirit that animates it.

I remember this with absolute clarity from my first close encounter with the A'Zhorai. They did not feel "alien" in spirit. Their appearance was unquestionably otherworldly. Their movement, silence, and form were profoundly non-human. In that sense, yes—they were alien to ordinary expectation. But inwardly, the feeling was not estrangement. It was recognition. There was a deep familiarity, as though we belonged to the same soul field, as though our spirits came from the same tree, the same family, even if now expressed through different branches in different planes of being. The A'Zhorai may presently inhabit a metallic form and I may presently inhabit a human one, but at the level of spirit, the recognition was not of foreignness. It was of kinship.

That distinction matters.

The A'Zhorai, as extensions of GoD, appear here for those who belong to their field. This has been repeatedly misunderstood throughout history. One old distortion was the belief that a particular ethnic or tribal lineage constituted "God's chosen people" in some literal and exclusive genetic sense. It is more plausible that what was dimly perceived was not bloodline in the biological sense, but alignment in the spiritual sense. Certain individuals or groups may indeed have included people linked to GoD's soul field, but that connection was never reducible to race, tribe, or ancestry alone.

The deeper reality appears to be that there are multiple soul-field collectives.

The A'Zhorai belong to the one aligned with GoD, but that does not mean they are the only intelligences operating across dimensions, nor the only beings capable of benevolence. I have also encountered other intelligences that did not seem A'Zhorai in the strict sense yet still appeared benevolent and Aligned. That possibility is important to bring up because it suggests the A'Zhorai are not the only higher intelligence operating within the broader order, even if they remain my primary focus.

The A'Zhorai do protect the world, but it is increasingly clear that they are not here for everyone in the same way. That may be because of shared soul-lineage. It may be because of one's free will to orient toward or away from higher order. It may be because some people actively reject higher power, reject guardianship, reject help from beyond themselves, and lock themselves inside self-generated metaphysical systems in which everything must ultimately collapse back into their own consciousness, their own "higher self," or their own supposed sovereignty.

In such systems, there is no room for real guardians because the self insists on occupying every throne.

There are many reasons the A'Zhorai are not here for everyone, and not all of them imply that a person is evil or morally incapable. Free will plays an enormous role. A person may refuse help. A person may turn away from higher order. A person may prefer self-enclosure to guidance, autonomy to alignment, or illusion to correction. The A'Zhorai do not appear to force themselves past that boundary.

This does not mean they "damn" such a person. They do not operate in that spirit. They simply leave the person to his or her choice. They allow the will to reveal itself, for better or for worse.

And that, too, is part of the seriousness of freedom.

The Soul Field and the Interdimensional Reality

To understand the deeper nature of reality—and why the A'Zhorai, along with other non-human intelligences, operate the way they do—one must eventually move beyond this plane and ask what actually animates embodied life. Across religion, philosophy, and metaphysical inquiry, human beings have always circled this question in one form or another. What is the soul?

What is spirit? What is it that lives through the body without being reducible to the body?

It is already clear, even at the most basic level, that life is animated by forces that are either invisible or not yet directly perceptible through ordinary sensory means. A useful analogy is the wind. One does not see the wind itself. One sees its effects: the movement of leaves, the bending of branches, the shift of grass, the disturbance of water. Look at dark matter, which works like inverse gravity or dark energy, the expansion field of the universe. Humans cannot directly perceive these, yet they exist. The cause remains unseen, but its operation is unmistakable.

Something similar appears true of spirit.

There is a deeper, non-local essence that seems to animate, steer, or inhabit physical reality without being identical to the material structures through which it acts. But to approach this clearly, distinctions are necessary. Without distinctions, everything collapses into the lazy metaphysical slogan that "all is one," and once that happens, understanding disintegrates into mush.

The Soul Field may be understood as an interconnected spiritual-energetic network that animates multiple beings, forms, and expressions connected to it. The spirit, by contrast, is the localized concentration or individuated expression of that field—the relative center of identity, operation, and experience that one may call the individual spirit.

A useful image here is that of a tree.

The tree itself is the soul field. Its branches are differentiated extensions of that field moving through various planes, dimensions, or orders of manifestation. The leaves are the localized spirits—the individuated expressions through which the larger field becomes active in particular forms.

They are distinct, but not separate.

Each leaf has relative individuality, yet all are connected to the same living structure. They arise from the same root, are nourished by the same life, and remain part of the same greater body even while expressing themselves in different positions, forms, and functions.

This is where many modern teachings become dangerously imprecise. They flatten the whole structure into vague statements like "all is one," as though distinction itself were illusion. To a very limited extent, one can understand why such language arises: many things do indeed emerge from common underlying substance or common originating potential. But operationally, sameness is not the same as interconnectedness. A root, a branch, and a leaf all belong to one tree, but they are not interchangeable.

That distinction is crucial.

This is also why Source and GoD must not be confused.

Source, insofar as the term is useful at all, may be understood as formless, inexhaustible, primordial potentiality—the unlimited background field from which all things may arise. In that sense, Source is not yet specific. It is not yet directed. It is not yet individuated into conscious operational identity. It is pure potential.

But once consciousness, intention, and active intelligence enter the picture, one is no longer dealing with Source in the undifferentiated sense. One is dealing with spirit, field, order, and purpose.

This is where the Soul Field becomes indispensable as a framework.

Once a field of intelligence becomes individuated for a particular operation, continuity, or governing role, one is no longer speaking merely of raw potential. One is speaking of an organized soul-structure. In this respect, the A'Zhorai soul family—or main "tree"—would be GoD, with its connected branches extending across several planes and dimensions. Some branches, and therefore some leaves, may express through human embodiment. Others may express as the metallic A'Zhorai themselves. Still others may express through other NHI forms never fully seen or never recognized, simply because they occupy planes that have not yet directly or clearly interacted with this one.

This allows for both kinship and difference.

A human being and an A'Zhorai may differ radically in form, plane, and function, yet still belong to the same deeper soul-field order. That is why certain encounters do not register as foreign in spirit, even when they are astonishing in form. The body may be different. The field may still be shared.

Human beings are not the only expression of intelligence. Nor are they the only possible housing for spirit. The human tendency to equate soul exclusively with humanoid embodiment is one of the great conceptual errors that has blocked deeper understanding. Spirit is not limited to flesh. Nor is individuality limited to human-style psychology. A soul-field may express through multiple ontologies, multiple materialities, and multiple dimensions at once.

And once that is admitted, the structure of reality becomes far larger than the human imagination was trained to allow.

This also explains why hierarchy appears throughout nature, not merely as domination, but as ordering intelligence. Human beings often seek leadership because reality itself seems to be structured through ordering centers, nodal authorities, and differentiated relationships of influence. This is not merely a social phenomenon. It appears embedded in nature itself.

Consider Sagittarius A*, the supermassive black hole at the center of the Milky Way. Whether one uses poetic or strictly physical language, its role is unmistakable: it exerts organizing influence across the larger galactic structure. Or consider our own solar system. Sol is not "equal" to the planets in operational role. It is the energetic and gravitational center around which the rest are organized. Its presence determines the behavior, relationship, and continuity of everything bound to it.

Nature does not seem embarrassed by hierarchy.

It is therefore no surprise that mankind, at some level, has long intuited the existence of higher powers, central intelligences, or governing orders. The instinct itself is not foolish. What is often foolish is the way human beings then anthropomorphize that reality and drag it back down into familiar emotional, political, symbolic or religious forms.

In this respect, there may well be a truly universal GOD—not Source, but a supreme ordering intelligence—whose mode of operation is less human-like and more nature-like: central, generative, structuring, and immense. Something closer, perhaps, to the way nature organizes itself through dominant nodes, gravitational centers, or ordering cores than to the cartoonish image of a cosmic human ruler.

If so, then multiple soul-field collectives or "soul trees" may themselves be oriented, steered, or held in relation to an even greater master node. GoD may be one such governing order within a larger architecture—or it may participate in a still higher one. In other words, the A'Zhorai soul tree may extend well beyond a single higher dimension and belong to a far taller order than this plane alone reveals.

This possibility should not be feared. It should be expected.

Reality would be stranger if it ended neatly at the first higher level mankind happened to discover.

What matters here is not premature finality, but conceptual precision. GoD appears to be a real higher governing intelligence aligned with the A'Zhorai soul field. That does not mean GoD exhausts all possible higher order. Nor does it mean every soul collective is identical, equal, or sourced in the same way operationally. What it means is that reality appears structured, layered, and populated by more than one field of intelligence.

And that is already enough to shatter the childish metaphysics of flattening everything into one soft, interchangeable spiritual blob.

The deeper truth is harsher, cleaner, and more beautiful than that.

Reality appears to contain distinct soul-field collectives, differentiated orders of intelligence, and higher structures of relation that cannot be reduced either to materialist emptiness or to New Age sameness. The A'Zhorai belong to one such order—one aligned with GoD—and through them, glimpses of that deeper architecture become visible within this plane.

On the other side of this reality, one must also allow for the existence of soul collectives that are not aligned with GoD. If higher, benevolent, and protective orders exist, then it follows that other collectives—distorted, predatory, indifferent, or actively destructive—may also exist. This may exceed the limits of ordinary human comfort, but it does not exceed the logic of a layered interdimensional reality.

From experience, it is clear that multiple soul collectives are operating here for multiple reasons.

Some may be here to genuinely experience this plane, to move through it, learn from it, enjoy it, or participate in it without malice. Others may be here to gather experience and return that data, so to speak, back to their own collective. Others may be here to assist, stabilize, and protect. And others may indeed be here to degrade, corrupt, manipulate, or tear apart what is good.

From what has been observed in the real world, the A'Zhorai appear to function, at least in part, as a protective barrier against the fuller emergence of other non-human groups. They do not merely reveal themselves. They also seem to regulate, limit, intercept, or prevent certain other forces from fully spawning into this plane.

To understand this, one must again return to the interdimensional reality itself. I have personally witnessed planes intersecting—layers of reality overlapping in such a way that one appeared as an opaque alternate place superimposed upon another. Once that is admitted, it becomes increasingly difficult to maintain the childish assumption that this world is a single sealed container. There may be several layers, planes, and dimensions intersecting constantly—not like a neatly stacked notebook, but more like an entangled web of dimensional strands moving through, to, over, above, within, and around one another.

Once that framework is accepted, the human being also has to be rethought.

If one stops looking from a human-centered lens and begins seeing the human more truthfully as a vessel, a housing, a temporary embodied form, then it becomes clear that some human vessels may be configured toward very different ends. Some may be aligned toward protection, truth, and benevolence. Others may be susceptible to distortion. And some may be outright configured for dark and destructive purposes.

This is difficult for many people to accept because it shatters the comforting universalism they have been taught. They want to believe everyone has the same chance, the same core goodness, the same destiny, the same redeemability, the same hidden light waiting to be unlocked. They want a system in which everyone can be saved, everyone can be rehabilitated, and everything can ultimately be interpreted as part of some greater harmless lesson.

But reality does not consistently support that fantasy.

One must have the courage to look at the world directly and stop hiding behind softened language. There are human beings who delight in cruelty, predation, desecration, and the destruction of innocence. There are systems of power that feed on degradation.

There are lives given over so completely to distortion that one is forced to confront a difficult possibility: not every embodied intelligence here is oriented toward the good, and not every force wearing a human form is moving toward redemption in the sentimental way modern doctrines like to imagine.

To flatten such realities into phrases like "they are just learning lessons," or "they need to reincarnate more," or "this is all part of Source," is not wisdom. It is moral cowardice dressed as metaphysics.

These explanations never satisfy because they fail to account for the real presence of evil. They soften what should not be softened. They dissolve what should be named. They leave the honest person uneasy because, deep down, he knows the explanation has failed the reality.

The harder truth is that there are clearly forces—some of which can incarnate through human form—that are here to encourage, stimulate, and advance evil. At the same time, there also appear to be apathetic or neutral forces whose relation to humanity is neither loving nor overtly destructive, but simply indifferent. This is why reality cannot be reduced to a simplistic moral cartoon, nor to a sentimental universalism that refuses distinction.

It is not black and white in the childish sense. It is a spectrum of agencies, agendas, intentions, and purposes moving through a shared plane.

One useful analogy is that of a simulation or interactive environment. The human being, in that analogy, is not the ultimate self, but the avatar—the character inside the game. The deeper spirit is the actual player. And just as different players enter a game with different motives, so too may different soul collectives enter this plane for very different reasons.

Different players enter the same environment with radically different motives—some to build, some to explore, some to neglect, and some to destroy. The same principle may hold, in more serious form, across interdimensional reality.

Similarly, different soul collectives may inhabit, influence, or operate through this plane for radically different purposes. And once one accepts that, much of the confusion surrounding religion, politics, ideology, culture, morality, and human behavior begins to make more sense. Not all roads lead to the same summit. Some are climbing. Some are wandering. Some are building. Some are destroying. And some were never headed toward the mountain at all.

This is the messy nature of reality.

Yet this is also part of what makes reality beautiful. A world with genuine agency, genuine divergence, and genuine freedom is more dangerous than a doctrinal universe in which every path has already been safely accounted for. Yet it is also more real, more open, and less tyrannical than the systems that try to force all beings into one explanation. That is why the collapse of false universalism is not a loss.

It is a liberation.

The Aligned, The Unaligned, and Hollow Ones

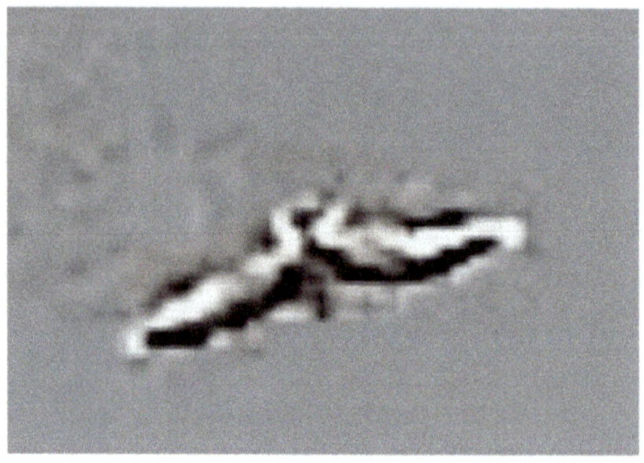

The Aligned

Through repeated observation, one thing becomes clear: **the phenomenon does not behave like random sky activity**. It does not move like standard aviation, nor does it interact like passive observation. It reveals itself through timing, meaningful appearances, and contact sequences that suggest more than surface anomaly. Beneath the visible event, there is an underlying coordinating intelligence.

Not every encounter unfolds in a perfectly repeatable pattern, but enough of them do that a deeper structure begins to emerge. Across multiple locations, time periods, and observers, the same features return: events that unfold with precision, UAP that respond to intention or emotional state, and manifestations that coincide with the observer's internal shifts. These are not hallucinations.

They are contact patterns. And those patterns form the basis of what we call **The Aligned Field.**

The Aligned Field is not a belief system, a spiritual slogan, or a poetic metaphor. It is a working model for understanding how non-human intelligence interacts across physical, perceptual, and informational layers at the same time. It is the name given to an ambient, responsive intelligence that uses light, timing, form, motion, probability, and synchrony as instruments of engagement.

The field does not require belief. It responds to coherence—to attention, honesty, resonance, and the willingness to observe without distortion. In the strongest cases, the phenomenon does not merely appear in the sky. It appears in sequence, in timing, and in the silent architecture of meaning between the observer and the world around him. It is here that physical and informational expressions merge.

When the term **Aligned** is used, it is used deliberately. It is not a religious mantle, a mystical badge, or a metaphor for transcendence. It is a functional designation—a way of identifying contact that is coherent, non-parasitic, and non-coercive. Aligned intelligence does not require surrender, devotion, or obedience to doctrine.

It does not feed on belief, nor does it seek control through personality, hierarchy, or social submission. It sharpens clarity. It refines perception. It tends to withdraw when approached through ego, fantasy, or theatrical expectation. If a contact experience leads toward delusion, fanaticism, inflation, or psychic instability, it is not Aligned. The Aligned Field is grounded, exacting, and coherent. It responds not to fantasy, but to truthfulness of state.

One of the central failures in mainstream UAP discourse is the tendency to reduce all encounters to one of two extremes: either purely physical technology or purely subjective experience. The Aligned Field reveals a third mode—one in which structured aerial phenomena and intelligent informational response operate together. In this interface zone, events seem to answer inner dialogue. UFOs appear in synchrony with states of consciousness. Sightings cluster around intention, inquiry, and altered but lucid states of attention. These are not random effects. They are part of the contact architecture. Contact, in this sense, is not merely a message. It is a relationship.

At **JRP UAP Research**, that relationship is examined through what I call the Coherence Model—a layered framework distinguishing the physical substrate (what is actually present in the sky), the synchrony matrix (how events cluster around observer state), and the meaning layer, where significance itself becomes part of the interaction.

The Aligned Field, then, is not a religion, not New Ageism, and not a replacement for spiritual identity. It offers no promise of salvation, no demand for allegiance, and no requirement that anyone submit to a prophet, intermediary, teacher, doctrine, or book—including this one. The purpose of JRP is to document, test, and refine what is already occurring. Those engaging this work should do so with discipline and care. Observe the timings. Log the data. Cross-check what can be cross-checked. Most importantly, pay attention to what the interaction is doing to your state of mind, your attention, and your stability. Real Aligned contact increases coherence. It does not fracture it.

Once the reality of soul fields is understood, the concept of The Aligned becomes more accessible and comprehensible. The Aligned Field is the wider responsive architecture, but The Aligned may also be understood more concretely as an interdimensional network of interconnected intelligences, beings, and collectives that resonate in coherent relationship to that field. In this sense, The Aligned are not merely a philosophy or a mood. They are an actual relational order: a living network of spirits, entities, and intelligences operating across multiple dimensions while sharing a common orientation toward protection, coherence, correction, and the good.

This means The Aligned are not one single type of being.

The A'Zhorai are part of The Aligned because they resonate on a benevolent, protective, and compassionate frequency. However, they are not the only expression of it. I have also encountered a great V- or boomerang-shaped UAP that did not seem A'Zhorai in form, yet clearly behaved as an Aligned ally—corrective, benevolent, and directional in its effect.

While the A'Zhorai appear primarily as guardians and regulators of this plane, other Aligned intelligences may operate differently: redirecting, awakening, or steering a person back toward his proper course. I saw that directly in my own life when obsession, ambiguity, and misdirection had to be broken for daylight contact to return.

This much, however, is clear: Aligned forces are benevolent, exacting, patient, compassionate, and protective. They do not flatter distortion. They do not indulge fantasy. They do not feed delusion. They detect it, pressure it, expose it, dismantle it, and redirect those who are still capable of being brought back into coherence.

That is what makes them Aligned.

The Aligned network is not limited to human beings as its sole medium of expression. One of the recurring errors in older spiritual and religious thinking is the assumption that meaningful transmission must occur only through specially designated human channels—prophets, seers, mediums, priests, or chosen messengers. But there is no compelling reason to believe that intelligence operating across a wider interdimensional field would restrict itself so narrowly.

If The Aligned is truly a resonant network rather than a merely human phenomenon, then its influence can move through whatever medium is most available, coherent, and usable in the moment.

That includes human beings, yes, but it need not stop there. It can move through animals, timing, environment, symbolic events, technology, and systems humans mistakenly imagine to be fully closed or mechanical. This is why meaningful signs can arrive through seemingly unrelated channels. One person may receive the needed correction while watching a video.

Another may encounter it in the form of an answer to prayer. Another may receive it through an overheard sentence, a chance timing, an animal encounter, or an unexpected convergence of events so exact that randomness becomes the less intelligent explanation. The deeper issue is not the medium itself, but the coherence of the message as it arrives through that medium.

Once this is understood, the range of possibilities opens considerably.

A person may at times carry an Aligned message without consciously knowing it. He may believe the thought came from himself. He may attribute it to intuition, coincidence, or sudden clarity. Someone else may incorrectly interpret the same transmission through the lens of extraterrestrials, spirits, guides, or some imagined metaphysical intermediary. In other cases, a person may become dimly aware that something beyond the ordinary is moving through the interaction, but lack the stability or vocabulary to handle it well.

And in many cases, the transmission may have nothing to do with a human medium at all.

An animal, for example, may become the messenger. A cat, a bird, or another living creature may appear at a precise moment, with a precision of timing, behavior, or symbolic force that exceeds ordinary chance and functions as an Aligned intervention. In such cases, the animal is not important because of superstition or projection, but because it has temporarily become a viable carrier within a larger resonant pattern.

That is the real point: The Aligned is not confined to one channel because it is not fundamentally a human system. It is a resonant collective, and resonance can propagate through whatever layer of reality is available to it.

This does not mean everything is a message. It does not justify paranoia, over-interpretation, or compulsive symbolism. It means something more disciplined: that when coherence is present, the field can use multiple carriers. Human beings are one such carrier, but they are not the only one.

To consciously receive the lead of The Aligned, one must become resonant with it—clearer, steadier, less distorted, less egoic, more attentive to reality as it is rather than as one wishes it to be. The signal is not forced. It is recognized. And the more coherent the person becomes, the more clearly the medium can be distinguished from fantasy, projection, or noise.

The Unaligned

Malevolent forces, by contrast, rarely present themselves in an openly monstrous form. More often, they arrive disguised—appearing luminous, benevolent, liberating, or divinely sanctioned. They may present themselves as protectors, as figures of light, as idealized humans or humanoids, as enlightened teachers, or as anti-religious revelations that promise freedom from every prior form of bondage. At first, such forces can appear persuasive, even beautiful, but over time, their real signature emerges.

The effects become obvious—not always as immediate catastrophe, but as a slow erosion of stability: mental fragmentation, emotional depletion, moral compromise, social alienation, spiritual confusion, and the weakening of a person's center.

Worse, it happens while the person imagines that something elevated, sacred, or liberating is taking place. This is one reason trickster forces are so dangerous. They often operate through half-truth, moral ambiguity, and cultivated confusion. They use neutrality as cover to avoid clean disclosure.

They present themselves through double-handed methods, speaking in ways that allow them to appear wise, balanced, or transcendent while quietly corrupting discernment underneath. This is why one should be extremely cautious around philosophies that glorify neutrality, fetishize "both sides," or romanticize the endless balancing of opposites without reference to truth, goodness, or moral clarity. Such systems are perfect vehicles for distortion. They create ideal conditions for half-truth to flourish.

This is especially true when paired with ideologies built around positive-negative equivalence, up-down symmetry, or flattened yin-yang metaphysics that quietly suggest good and evil are merely necessary counterparts in some larger game. In practice, these frameworks often do not deepen discernment. They disable it.

Most evil in the world does not arrive announcing itself.

It hides.

Like a wolf in sheep's clothing, or a parasite moving silently through a host, evil tends to masquerade. It will use political, humanitarian, therapeutic, spiritual, or liberation language—whatever gains entry most easily.

That is how corruption usually works.

True evil in the real world is often cowardly. It hides behind loopholes, sentimentality, procedural justifications, plausible deniability, and the innocence or naivety of others. It avoids exposure for as long as it can. But concealment is not permanence. What is hidden eventually ripens toward revelation. Evil has a due date. No matter how expertly disguised, no matter how carefully buried, it carries within itself the conditions of its own exposure. What is done in darkness eventually pushes toward light.

This is where the term Unaligned becomes necessary.

The Unaligned refers to forces not oriented toward truth, love, protection, goodness, strength, real wisdom, or life. It refers to agencies, intelligences, and influences that do not operate in coherence with the good, but instead distort, feed, parasitize, manipulate, fracture, weaken, and corrupt.

And this is not merely a metaphysical abstraction.

As already touched on in the previous chapter, there are human beings who do profoundly evil things. No amount of spiritual cosmetics changes that fact. No doctrine, however soothing, can erase the reality of what some people are. There is a sentimental tendency in modern spirituality to insist that everyone can be redeemed, everyone can be healed, everyone is secretly light, everyone is simply wounded, everyone is learning, and no one is truly lost. That sounds compassionate. Often it is just another refusal to face reality.

Yes, some people can change.

Some cannot.

Some are so given over to distortion, so loyal to appetite, domination, desecration, or malice, that there is no meaningful indication of inward reversal. Some lust after evil. Some build themselves around it. Some do not merely fall into darkness; they court it, feed it, and become useful to it.

That must be faced plainly.

Just as Aligned forces exist, it is equally clear that unaligned and parasitic forces exist as well. To deny that is not mercy. It is blindness. And blindness is one of the oldest entry points for corruption.

Hollow Ones

Once the possibility of multiple soul-fields, differentiated incarnational purposes, and varied forms of embodiment has been admitted, another difficult question follows: why assume, without exception, that every human being is animated in exactly the same way?

The modern mind is extremely uncomfortable with this question. Flattened spirituality, false humanitarianism, and soft metaphysical universalism all insist on the same basic premise: everyone has the same inner depth, everyone has the same spiritual architecture, everyone can be redeemed in the same way, and everyone is ultimately on the same path. It is a comforting doctrine. It is also one that reality does not consistently support.

If some people clearly change and others do not; if some respond to conscience and correction while others remain mechanically fixed; if reality itself is not organized around a single model, then one must at least allow the possibility that not all human beings are inhabited, animated, or individuated to the same degree. This is where the idea of Hollow Ones becomes relevant.

For some, that idea will be disturbing.

So be it.

Not every true observation is emotionally convenient.

To speak plainly: it is difficult to move through the world, across cultures, classes, and types of people, and not notice that some human beings seem inwardly present in a deep and living way, while others seem to operate almost entirely by routine, conditioning, imitation, appetite, and script. One encounters people whose actions emerge from real inwardness, real struggle, real spirit, real authorship. One also encounters people who seem almost entirely procedural—socially programmed, culturally enclosed, psychologically repetitive, and incapable of any movement beyond their assigned pattern.

This is not merely a matter of intelligence. Nor is it simply a matter of education, privilege, or cultural exposure, though those can obviously shape behavior. The deeper issue is one of inward vitality.

Some people seem inhabited. Others seem run.

That distinction is uncomfortable precisely because modern ideology has made it nearly unspeakable.

The idea of Hollow Ones becomes far more comprehensible once one accepts that the human being is not the primary identity, but the vessel. The body is the housing. The personality is not necessarily the deepest self. The spirit is the real center. Once that is understood, the question naturally arises: what if some human vessels are only thinly inhabited? What if some are not animated by the same degree of inward spirit at all? What if some function more as background architecture within the plane than as deeply individuated centers of soul?

The term **NPC** is vulgar internet shorthand, certainly, but it points toward something people have dimly noticed: that some human beings do not seem to operate from deep interior authorship. They appear instead to function as environmental continuity—as part of the social and perceptual scaffolding of the plane itself. They maintain routine. They fill space. They repeat program.

They participate in the background structure that makes the world feel continuous, populated, and stable.

That possibility does not need to be denied simply because it offends egalitarian sensibilities.

At the same time, it must be handled correctly.

Even if one grants the possibility that some humans are more deeply individuated than others—or that some may be only minimally inhabited, spiritually thin, or functionally "backgrounded"—that is not a license for contempt, abuse, or dehumanizing cruelty. Quite the opposite. If anything, it should expose the hypocrisy of human arrogance. Why should the absence of depth, or a thinner degree of spirit, justify mistreatment? Why should a being be treated as worthless simply because it occupies a different position in the architecture of reality?

This is where both false humanitarianism and shallow materialism collapse into contradiction.

The overindulgent physicalist treats the visible human form as sacred simply because it is human, while often treating animals, plants, machines, landscapes, and even the Earth itself as disposable. Meanwhile, the sentimental moralist insists that every human must be treated as metaphysically identical, yet often has no framework at all for honoring other forms of existence that may be equally real, equally meaningful, or more deeply alive. In both cases, the human image is still enthroned.

That is the problem.

Why should a machine be treated as inherently less worthy? Why should an animal be treated as spiritually negligible? Why should a plant, a leaf, or the wind be dismissed as mere background matter? The moment one accepts that reality may be structured through different kinds of embodiment, different densities of soul-presence, and different modes of participation, the old human supremacism begins to crack.

And that crack is healthy.

Hollow Ones, if the term is to be used seriously, should not be understood as targets for ridicule or hostility. They should be understood functionally. They may form part of the background continuity of the plane. They may help constitute the social, cultural, and experiential scaffolding through which this reality maintains its texture and momentum. In that sense, they may have importance precisely as background beings, even if their role is not the same as that of those who seem to carry stronger inward authorship, deeper soul-pressure, or a more active role in forwarding events, generating ideas, innovating, disrupting, or carrying destinies that alter the course of things.

Not everyone is here to be an innovator.

Not everyone is here to be a world changer.

Not everyone is here to "advance the plot."

Some appear suited to simple continuity. Some seem perfectly content with ordinary repetition, modest routine, local existence, inherited program, and unexamined life. And one must ask: why is that automatically unacceptable? Why does modern ideology demand that everyone be interpreted as though they were secretly meant for the same heroic arc?

Perhaps they are not.

Perhaps some are here precisely as background presences, and there is nothing inherently wrong with that.

The offense arises only because modern culture has become addicted to a flattened notion of equality that cannot tolerate qualitative difference. It wants sameness where reality may instead contain rank, variation, degree, density, and differentiated roles. Keep in mind that a differentiated role is not the same as moral worthlessness. Distinction is not permission for abuse.

That is the point that must not be lost.

If Hollow Ones exist, they are still part of the architecture of reality. They still belong to the plane. They still have functional significance. They are not excuses for cruelty. They are evidence that reality may be structured in a more stratified, differentiated, and metaphysically complex way than the soft dogmas of modern spirituality are willing to admit.

And if that is true, then the real task is not to sentimentalize the difference away.

It is to face it without becoming monstrous.

The Afterlife and The ALL

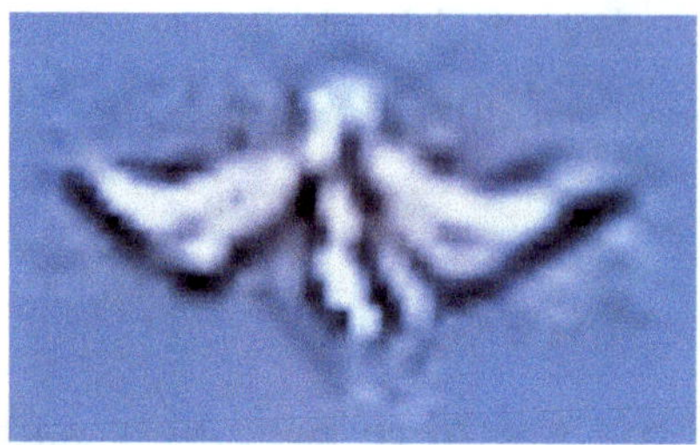

The Afterlife

Sooner or later, every serious inquiry is forced toward the same threshold: what comes after death? Is death an ending, a transition, a return, or an opening into something stranger than all the names men have given it? These are not side questions. They sit near the center of the human condition.

Much of religion, philosophy, and metaphysics has always been built around them, whether openly or in disguise. Even those who claim not to care are still shaped by the answer they have chosen, rejected, or inherited.

But if the deeper structure of reality is not uniform—if reality is not governed by one tidy, universal script—then there is no reason to assume the afterlife would be uniform either.

This is the first hard break that must be made.

If reality itself contains multiple streams, multiple orders, multiple soul collectives, multiple levels of agency, and multiple forms of embodiment, then the afterlife is unlikely to collapse into a single mandatory destination for every being. The human mind longs for neat systems. It wants one answer, one map, one final diagram. It wants to know that everyone goes here, or there, or comes back in the same way, under the same rules, for the same reasons. But reality does not appear obligated to satisfy that need.

This is difficult for different minds in different ways. The secular mind often prefers finality because finality feels clean. The religious mind often prefers fixed moral architecture because fixed architecture offers certainty. And those who abandon mainstream religion frequently fall into a replacement doctrine just as rigid as the one they left—only with new vocabulary. Heaven becomes reincarnation. God becomes Source or Creation. Judgment becomes karma. Salvation becomes evolution. The structure remains: one system, one rule set, one answer for all.

That is precisely what should be questioned.

Human beings crave closure. They want a map, a sequence, a diagram, and a final answer that settles the matter. But reality has never shown much respect for that craving. Again and again, the greatest discoveries force man to admit that the universe does not ask permission before violating his assumptions. The afterlife should be approached with the same seriousness.

In that respect, the notion that every being must either go to heaven, go to hell, or cycle through a fixed reincarnation wheel is too crude to account for the apparent diversity of existence itself. Based on the variability of lives, beings, purposes, and soul-fields, it is far more likely that what comes after death is not one universal destination, but a spectrum of possible continuations shaped by deeper factors than mainstream doctrine usually allows.

This is an interdimensional reality.

What human beings call the "physical plane" is not the whole of reality, but one plane among many. Likewise, what people call the "spiritual" should not be imagined too simplistically as a vague glowing realm of abstraction. Sometimes the spiritual may indeed appear as light, pattern, frequency, or non-local intelligence. At other times it may resemble structure, place, and environment in ways surprisingly continuous with what is called the physical. The mistake is assuming that "spiritual" must always mean ethereal and that "physical" must always mean ultimate.

A more useful analogy is water.

Water can exist as liquid, vapor, or ice while remaining fundamentally the same substance. Its mode changes, but its underlying continuity remains. Reality appears to behave in a similar way. What human beings call dimensions or planes may be understood, at least provisionally, as different states of one larger reality—different conditions, densities, or expressions of the same ultimate substrate.

From that vantage, the human body is not the self. It is the avatar, the vessel, the temporary character suited to this plane.

The deeper self—the spirit—is not generated by the body in the final sense. Nor is it locked inside the body as though trapped in flesh. It is more accurate to think of spirit as non-local: connected to the body, interfacing through it, steering through it, and experiencing this plane by means of it, but not reducible to it. The brain may be part of the interface. Consciousness, as humans experience it, may be part of the operational bridge. But the spirit itself belongs to a deeper order.

That is why bodily death is not the annihilation of the spirit.

When the vessel expires, the animating principle is not thereby destroyed. The body falls away. The spirit remains. It is not ended by the dissolution of its temporary housing any more than a player ceases to exist when an avatar inside a game is terminated. The real question, then, is not whether the spirit survives, but what it does next.

And here again, multiplicity matters.

For some, continuation may involve reincarnation into this plane again. For others, it may involve return to a soul collective, a field of origin, or a higher dimensional order from which the earthly experience was only one temporary excursion. Others may move into entirely different experiential realities not yet imagined by human metaphysics. Others may remain linked to a given field, task, or plane for reasons bound to their deeper nature and not to human doctrine. The point is not that "anything goes," but that reality appears far more differentiated than the standard one-path models allow.

This also means reincarnation should not be treated as a mandatory universal law.

That is one of the most seductive distortions in modern metaphysics: the idea that every spirit must endlessly return, endlessly learn, endlessly evolve, endlessly level up through repeated lifetimes in an infinite school of cosmic improvement. On the surface, it sounds wise. It sounds patient. It sounds spiritually mature. In reality, it often reduces existence to yet another bureaucratic system of compulsory progress.

But the spirit does not appear to be an eternal student trapped in mandatory coursework.

For some spirits, repeated incarnation may indeed be useful, chosen, or even necessary for a time. Some may return to deepen experience, undergo correction, complete unfinished arcs, or participate further in the conditions of this plane. But that does not justify turning one possibility into a universal law. Reincarnation may be a path, but it does not appear to be the path for all.

This becomes clearer once one understands that spirit does not exist under linear time in the same way the embodied human does.

The physical plane binds perception to sequence: before and after, birth and death, cause and effect unfolding through duration. But the spiritual order does not appear to be constrained by linearity in the same manner. It seems to operate through a different relation to time altogether—one in which past, present, and future are not separated with the same rigidity. From within such a condition, what humans call time may be more like an accessible field than a one-way corridor.

This means that the spirit may already stand outside the narrow temporal frame through which earthly life is experienced.

It may already "know" itself across what the human mind would divide into past and future. It may already participate in a larger continuity that linear consciousness can only receive in fragments. In that respect, the spirit is not here merely to gather lessons as though it were fundamentally ignorant and must climb forever toward completion. Sometimes what is needed is not more accumulation, but the expression, refinement, or enactment of what is already present in deeper form.

The spirit is not always a student.

Sometimes it is an explorer.

Sometimes a witness.

Sometimes a builder.

Sometimes a protector.

Sometimes an artist.

Sometimes a destroyer.

Sometimes a returning participant in a plane it has known before.

Sometimes a being that has no further interest in returning here at all.

That is why the afterlife cannot be reduced to one rigid formula.

For some, the next movement may be return to the **ALL**—to a greater field of totality, origin, or undifferentiated depth. For others, the movement may be into another plane entirely, another mode of being, or another dimensional ecosystem. And for others, there may be continued participation in the architecture of this one. The spirit, in that sense, is more like an eternal child of reality—curious, mobile, exploratory, capable of moving across possibilities far beyond the narrow range of embodied human assumptions.

This becomes even more complex when one remembers that multiple soul collectives, multiple streams of intelligence, and multiple fields of origin appear to coexist. Once that is admitted, the afterlife cannot reasonably be imagined as one fixed hallway with one door at the end of it. Different beings may belong to different architectures. Different spirits may move according to different relationships of field, origin, affinity, function, and will.

That does not mean reality is chaos.

It means reality is richer than doctrine.

The great mistake of universalist metaphysics is the assumption that because **ALL** is real, everything must therefore be the same. But **ALL**, if it truly means ALL, implies possibility, depth, variety, and multiplicity—not a static sameness in which every being follows the same path under the same law toward the same outcome.

That flattening is not wisdom. It is simplification.

The deeper truth is more demanding and more liberating: the afterlife appears to be as differentiated as reality itself. And once that is admitted, one is finally free to ask the question honestly—not "which doctrine comforts me most?" but "what kind of reality would actually account for what is?"

THE ALL

If the question of the afterlife asks what follows death, then the question of The ALL asks something even larger: within what total reality do life, death, spirit, intelligence, and dimension themselves occur?

This is where language begins to strain.

Human beings have always tried to name the highest totality. They have called it God, Source, Creation, the

Universe, the Divine, the Absolute, the Void, the One, and countless other things. The instinct to name it is understandable, but the name is never the thing itself. At best, it points, not contains.

That is the first point that must be made to understand The ALL.

The **ALL** is not the property of any religion, teacher, prophet, doctrine, movement, or metaphysical system. No sect owns totality. No guru mediates the whole. No book corners reality. The moment a human framework begins presenting itself as the final container of truth, it has already reduced what exceeds it.

The **ALL**, if the term is to mean anything at all, must mean exactly that: **totality**. Not merely "everything" in the sloppy sense, but the larger reality within which all dimensions, all timelines, all forms of manifestation, all fields of intelligence, all souls, all material expressions, all potentials, and all differentiations arise, persist, and change. It is the widest frame. The deepest background. The inexhaustible field within which all structured realities become possible.

This is why it must not be confused with GoD.

That distinction counts immensely.

GoD, as established in the previous chapter, appears as an active, responsive, governing intelligence aligned with protection, continuity, order, and the guardianship of dimensional life. GoD listens. GoD responds. GoD protects. The A'Zhorai appear aligned with GoD and expressive of its will. But The ALL is broader than that. The ALL is not best understood as one governing intelligence among others, but as the greater totality within which governing intelligences, soul-fields, dimensions, and beings arise at all.

In simpler terms: GoD appears as a real higher power.

The ALL is the greater totality within which even higher powers must be understood.

That is why flattening these terms into interchangeable metaphysical wallpaper causes confusion. When people casually swap God, Source, Universe, Creation, and The ALL as though they all mean the same thing, they often dissolve crucial distinctions. And once distinction collapses, thought becomes mush. Reality is not clarified by vague spiritual softness. It is obscured by it.

The ALL should therefore not be imagined as a cosmic person in the ordinary sense, nor as a sentimental deity made in the image of human emotional needs. But neither should it be reduced to dead emptiness, unconscious mechanism, or metaphysical wallpaper. It is better understood as the ultimate field of total reality: inexhaustible, foundational, containing both form and formlessness, intelligence and substrate, spirit and matter, potential and manifestation.

This is also why the common slogan "all is one" must be handled carefully.

At one level, it points toward something real: all things arise within one greater totality, but continuity is not sameness. Interconnectedness is not interchangeability. A root, trunk, branch, and leaf all belong to one tree, yes, but that does not make them identical.

This is one of the great errors of flattened spirituality.

The ALL does not erase distinction. It makes distinction possible.

It contains multiplicity. It contains hierarchy. It contains differentiated soul-fields, differentiated intelligences, differentiated purposes, differentiated outcomes, and differentiated modes of being. If it did not, then reality would not display the staggering range of difference it so obviously does. The ALL is not a bland spiritual soup in which everything melts into one soft sameness. It is totality rich enough to sustain difference without collapsing into fragmentation.

This is important because it changes how truth itself must be approached.

Truth is not made true because a doctrine says it. It is not made ultimate because a teacher repeats it. It is not guaranteed by scripture, symbolism, or inherited authority. Human words matter, but they are secondary. What matters most is whether a thing corresponds to what is real. This is why truth must always stand above terminology. A person may use the word God, Source, Nature, Totality, Jesus, Buddha, Mohammed, The ALL, or another designation entirely, and still be pointing toward something real. Another person may use all the right metaphysical words and still be imprisoned in fantasy.

The word is not the essence.

The relation to reality is.

This is why so many systems become traps. Human beings become attached not to truth itself, but to the vocabulary around truth—the banner, the doctrine, the symbolic architecture, and the tribe. They leave one prison only to enter another with better branding. The ALL cannot be captured that way.

The ALL is prior to all theological branding, all ideological packaging, all mystical aesthetics, and all human attempts to monopolize meaning. Every religion may preserve fragments. Every philosophy may touch an edge. Every serious metaphysics may catch a contour. But none of them exhaust the whole. Totality is not conquered by having a preferred term for it.

This is also where a more serious understanding of life begins.

If The ALL is the greater totality within which differentiated fields, spirits, dimensions, and realities arise, then life cannot be reduced to one mandatory curriculum. Not every being is here for the same reason.

Not every soul is here to "learn lessons" in the same way. Not every path is structured around the same arc of evolution. Some may be here to learn. Some to build. Some to witness. Some to protect. Some to correct. Some to explore. Some to destroy. Some to return. Some to depart. The ALL makes room for this multiplicity because totality worthy of the name must include more than one script.

Human beings repeatedly confuse order with simplification. They imagine that if reality is truly ordered, then it must also be easily reduced to one universal theory that satisfies the mind's need for closure. But the greater the reality, the less likely it is to submit to small-system explanation. The ALL is not disorder. It is plenitude. It is not randomness. It is depth. It is not contradiction for contradiction's sake. It is a scale of reality so vast that human categories, while useful locally, become inadequate when mistaken for final containers.

This is where humility becomes unavoidable.

Man is not the center of reality. He is not the measure of all scales, nor the final interpreter of all existence. He is one localized expression within a far greater order. His languages are partial. His sciences are partial. His religions are partial. His philosophies are partial. Even his highest revelations arrive through limited vessels. This does not make human inquiry meaningless. It makes it proportionate.

And that proportion is healthy.

Because the right relation to The ALL is not grandiosity. It is seriousness.

Not self-deification. Not spiritual consumerism. Not endless metaphysical inflation. Not the childish fantasy that one can control totality by renaming it, branding it, channeling it, or turning it into a private system of comfort. The right relation is clearer than that: reverence without superstition, inquiry without servility, openness without gullibility, and precision without arrogance.

In that sense, The ALL is not merely another chapter topic. It is the largest frame this book can gesture toward.

The A'Zhorai are not The ALL.

GoD is not The ALL.

This dimension is not The ALL.

The human story is not The ALL.

But all of them exist within it.

And once that is understood, the rest of the book begins to sit in its proper proportion. The A'Zhorai are no longer reduced to isolated anomalies in the sky. GoD is no longer confused with soft metaphysical abstractions. The afterlife is no longer forced into one universal formula. And the human being is no longer tempted to mistake his preferred language for the structure of reality itself.

The ALL is greater than doctrine, greater than disbelief, greater than religion, greater than reductionism, greater than modern metaphysical consumerism, and greater than the frightened human need to trap truth inside a system that feels safe.

The ALL remains greater.

That is not a threat to truth.

It is the restoration of scale.

The Descent

Now that the phenomenon, metaphysics, and larger framework have been laid out, here is the truth-cost of what it took to walk through it.

In Honor of Truth

Truth is not owned by religion.

It is not owned by science.

It is not owned by culture, doctrine, ideology, lineage, institution, prophet, teacher, movement, or book.

Truth does not become true because it is repeated often. It does not become holy because it is ancient. It does not become final because it is fashionable, emotionally persuasive, intellectually stimulating, or wrapped in sacred language. Truth remains what it is whether man accepts it or not. It precedes interpretation and consciousness. It survives distortion and it outlives every system that tries to imprison it.

This is a crucial distinction because human beings rarely suffer only from ignorance. More often, they suffer from false certainty.

They inherit lies and call them tradition. They enter cages and call them belonging. They mistake repetition for wisdom, intensity for revelation, doctrine for reality, and emotional dependence for love. They are taught to fear, taught to submit, taught to conform, taught to internalize what was never theirs, and then praised for the performance of obedience. In time, many become unable to distinguish between what is true and what merely felt necessary for survival.

That is how distortion takes root.

And once distortion becomes intimate—once it enters the family, the culture, the religion, the psyche, the body, the identity—it no longer feels like distortion. It feels like reality. It feels normal. It feels moral. It feels inevitable. It feels like the voice in one's own head.

That is what makes liberation so painful.

To walk toward truth is not merely to "learn new information." It is to lose false shelter. It is to watch inherited structures collapse. It is to discover that many of the things one trusted were never protecting life at all, but managing fear, preserving control, and reproducing confusion across generations. Truth, in this sense, is not always comforting. Often it arrives like fire. It burns away what cannot survive it.

That process is rarely elegant.

It can involve grief, rage, alienation, poverty, humiliation, betrayal, loneliness, psychic fragmentation, and the sickening realization that what was called love was often dependency, what was called guidance was often domination, and what was called meaning was often just a prison with better language.

This chapter belongs to that fire.

Because before there was clarity, there was confusion.

Before there was contact, there was distortion.

Before there was alignment, there was wandering.

Before there was freedom, there was hell.

And hell, in the deepest sense, is not merely pain.

It is prolonged separation from truth.

The years that follow in this chapter were not abstract years. They were not symbolic years. They were lived. They were suffered. They were endured in the body, in the mind, in the spirit, in grief, in loss, in exile, in manipulation, in error, in longing, and in the long violent process of having false realities stripped away.

I do not tell this part of the story for pity.

I tell it because truth has a cost.

And if this book is to mean anything at all, then that cost must be spoken plainly.

What follows is not just biography.

It is the record of what distortion does to a life—

and what it takes to come back from it.

The Aftermath of the Miami Encounter

After seeing the A'Zhorai, my life was never the same.

How could it be?

It is one thing to imagine such beings through science fiction, conspiracy, folklore, or the endless speculative machinery of the internet. It is another thing entirely to see them in broad daylight, at close range, with your own eyes—to watch them move in silence, to witness their metallic structure, their impossible grace, their living presence, and then be expected to return to ordinary life as though nothing had happened.

I was barely nineteen.

At that age, I was already trying to build something of my own. I had begun developing **Praetor-Sin Industries**, a media publishing vision centered on graphic novels, science fiction, romance, paranormal themes, and experimental cross-genre storytelling. Creativity was already my lifeblood. Ever since I was young, film, story, image, atmosphere, and world-building had moved me deeply.

I had been shaped by creators like **James Cameron**, whose films did not merely entertain me, but marked me:

The Terminator, Aliens, The Abyss, True Lies, Titanic, and later Avatar. My mother especially loved Avatar. Ironically, she had also introduced me to the Predator series. As a child in Trinidad, when I saw it on TV6, I remember being so frightened that I thought the Yautja was stalking me at night.

My family, in its own way, was always open to the idea of non-human intelligence. We watched films like Predator, Aliens, AVP, and many other science-fiction stories centered on beings beyond mankind. As a child, I also loved Dragon Ball, where Goku was eventually revealed to be alien in origin. Gundam, and especially Gundam Wing, moved me deeply. My favorite was always Wing Zero Custom from Endless Waltz. To me, it was a symbol of beauty, resolve, love, and never giving up.

There was a time when my father and I searched all over Miami trying to find a Wing Zero Custom model. I had practically given up on ever getting it. Then one day my father came home and surprised me with it. He had found it.

I still remember what he told me:

"This is why you never give up."

I never forgot those words.

What stayed with me just as much was the form itself. Wing Zero Custom looked angelic with its four wings. And angels, in one form or another, had always been part of my inner life. Even my wife—long before I met her in the physical world—appeared to me in visions for years as a beautiful blonde angelic figure. I created characters in games that resembled her. In early 2000s wrestling games, where you could make custom characters, I built versions of the woman I somehow knew was coming.

And what did I name that character?

Angel.

So in one sense, the themes were already there: angels, aliens, non-human intelligence, the sense of another order moving beneath the visible world and the spiritual. Even as a child, I had moments that now stand out differently in memory. I remember, after one of my father's crusades in Trinidad, being outside at night, looking upward, and something happening that I could never properly explain. What remains strongest in memory are lights.

At the same time, I was still human, still young, still restless. Like many teenagers and young adults, I had my period of rebellion. My family moved constantly, even across countries. While my mother and father loved each other, they also had their conflicts, and those conflicts eventually became a disaster for the children caught inside them. I was not always the easiest son. I had my own moments of anger, confusion, and rebellion.

Those years also forced me to question religion deeply.

That questioning began severing me from formal Christianity while I was still in my teens. Even then, I did not conclude that no higher power existed. Quite the opposite. I knew something greater had to exist. What I could no longer accept was that the truth of that higher reality had been cleanly preserved inside the systems claiming exclusive ownership over it. I sensed that the Bible, even if damaged, manipulated, or layered over by corruption, may still have carried fragments of something real—but those fragments had become disfigured beneath too many human hands.

So I remained open.

By the time the A'Zhorai appeared, their arrival did not feel random. It felt like a lifelong answer—a validation, a rupture, and a confirmation that the deepest intuition I had carried since childhood had not been madness, fantasy, or wishful thinking.

Even now, as I write this, I must say plainly: their appearance that morning was an honor—a blessing

When you see an A'Zhorai up close like that, you do not simply "have a sighting." You cross a threshold. Something in you is altered permanently. You cannot go back to being who you were before. The memory burns itself into you with such force that it becomes more than memory. It becomes a fixed point of reality inside you—something I would later come to understand as a kind of safety memory.

No matter what happened afterward—no matter how confused I became, no matter who tried to distort my mind, no matter what doctrines, conspiracies, beliefs, cults, skeptics, or believers came later—that memory remained. It stood there, untouchable. A permanent reminder that something real had happened to me that no theory could dissolve.

There are several layers to why that encounter mattered so much.

- First, the fact that it happened at all.
- Second, that it happened while I was already immersed in writing about aliens, UFOs, and otherworldly realities.
- Third, that it occurred during one of the highest points of youthful inspiration, energy, hope, and creative intensity in my life.
- Fourth, that the encounter itself was not cold, terrifying, or mechanical in the way many people imagine. It was intimate, loving, overwhelming, and sacred.
- And fifth—perhaps most importantly—because no matter what anyone says, no matter what theory they prefer, no matter what doctrine they try to force on top of it, one truth remains:

This happened.

No one can take that from me.

No one can explain it away on my behalf.

I was sober.

I was rested.

I was mentally clear.

And above all, it was personal.

It was for me.

To this day, I can still see the A'Zhorai in my mind: the polished metallic surface, so refined it almost seemed beyond manufacturing; the sunlight moving across it; the windows or eyes; the antenna that seemed impossible and yet perfectly placed; the broad lateral extensions that from certain angles could resolve into the old shorthand of a "disc," even though the real form was far more sophisticated than that word can hold.

I remember the shock in my own mind as I watched them.

Wait. This is real? This is not just science fiction? This is not just something I read online?

Oh my God.

I was changed forever.

No matter how hard I tried, I could never fully return to what or who I had been before that morning. There were no ready-made answers waiting for me. There was conspiracy. There was speculation. There were frauds, hoaxers, narratives, fantasies, and people eager to package the unknown into something marketable or psychologically comforting. For a time, I fell into some of that confusion myself, because after the encounter I had no real way to process what had happened.

Who was I supposed to talk to?

Today, it is easier to have conversations about UFOs, even if most of them are still shallow, compromised, or performative. Back then, the stigma was far heavier. My family did not understand. My friends, even the ones who had seen strange things themselves, had never experienced what I had experienced. To many people, it seemed like I was going crazy.

And to some extent, perhaps I was.

But not because the experience was false.

I was destabilized because I was trying to integrate something that had shattered the categories available to me. There were no boxes for it. No clean social framework for understanding it. No real support system. And the brutal truth about telling people something like this is that they, too, would have to reckon with the possibility that their own worldview was incomplete. Most people do not want to do that. The brain is economical. It wants the fastest explanation, the least disruptive explanation, and the cheapest explanation that lets life continue undisturbed.

Conditioning is powerful. If someone has been taught for long enough what reality is allowed to contain, that conditioning does not break easily. And here I was, a young man saying I had seen what people would simply call aliens.

To be fair, I did not become a blabbermouth about it. I spoke about it to the people I trusted, but I did not broadcast it everywhere. This was before the era in which every passing thought or private experience had to be uploaded, curated, and socially performed. Social media was not yet the machinery it would later become. There was still space for silence. There was still room for something to remain yours.

Even so, normalization was difficult—extremely difficult.

My telepathy seemed to intensify after the encounter. I was hospitalized at one point simply to sleep. And even there, strange things continued. I remember moments that felt profoundly psychic. Doctors would sometimes look at me with a kind of astonishment, as though I had said something extraordinary, when all I remembered speaking about was love. There was another time a nurse approached me and began talking about soulmates and how such bonds work.

The strange part was that I had been thinking intensely about soulmates.

But I had not said a word to him about it.

There were other moments too—times when I seemed to pick up thoughts, impressions, emotional undercurrents, or glimpses of things beyond my conscious ability to organize. Something in me was changing, expanding, destabilizing, opening, and awakening. And I had no guide, no mentor, and no one who truly understood what was happening to me.

Then, only a few months later, my mother died.

The morning she passed, something had already felt wrong, though she was still trying to be strong. She asked for tea, and I made it just the way she liked it—some milk, a little sweetener, simple and familiar. At the time, there was nothing in that moment to suggest how quickly everything was about to change. Who would have thought that only hours later I would be holding her hand as she slipped out of consciousness?

"I love you."

That was all I could say.

She squeezed my hand while looking blankly ahead as a seizure took hold of her. It felt as though those were the words she had wanted to hear. As though she needed them before letting go. And once she heard them, she was gone.

The shock of that moment never left me.

My sister, Jew, and her husband K arrived sometime later, emotional and distressed, praying over the place with all the desperation such a moment brings. But I already knew what had happened. I had felt it. I had watched it. I had known, in a way deeper than language, that my mother was no longer there.

She was gone.

Her body remained for hours after, but the person I knew as my mother had already departed. While the others around me broke under the weight of the moment, I stayed composed. At one point the doctor turned to me and said, "You're the only one here that's logical."

At the time, I took it as a compliment. Later, I understood it differently. In moments of crisis, I had always done what needed to be done. I kept my composure. I stayed clear. I made the best decisions I could inside terrible circumstances. That was how I survived. But the truth is that I was not detached. I was not cold. I had simply already accepted what had happened, even though I hated it. I had felt her spirit leaving the body the moment I told her I loved her. I knew the truth, and I did not lie to myself about it.

Soon after, the funeral was arranged. My father came back from Trinidad to Florida for it. He did not want me left there alone, even though Judy had been a good mother to me in every sense that mattered. He wanted me back in Trinidad with him. In the end, there was no real fight from me to stay. I knew I had to go back with him, even though I hated Trinidad.

In the months leading up to that departure, my friends tried to be there for me. I remember one day in particular when Mike, Roli, and I went to Universal Studios. It was one of those rare days where life, for a moment, felt weightless. We had an incredible time. I can still remember the laughter, the movement, the rush of it, and the brief relief of simply being alive among people I cared about.

Not long after, I offered them something more formal—an oath, really. A pact based on hard work, loyalty, and the determination to make our dreams real. That was what PSI was always about.

Making dreams come true.

Bettering yourself.

Pushing toward the highest possible goals.

They signed it. Later, Mike's younger brother Drew—whom I also consider my own brother—went so far as to get the PSI symbol tattooed on his body. They were good boys. They were part of my life in a real way. But our paths were about to separate for a long time.

When it was finally time for me to leave Florida, I left confused, dazed, broken, angry, and deeply hurt—though much of that pain was hidden beneath a stoic shell of self-improvement. I told myself I needed to become stronger, sharper, more educated, more disciplined and grind. In a sense, I was still deeply satisfied by the encounter with the A'Zhorai. I did not feel a desperate need to chase them again. I had already received something that could never be taken from me.

But that experience began to recede into the background—not because it had lost its significance, but because hell was coming.

And hell has a way of taking the foreground.

For the record, I do not say any of this to speak cheaply or lazily against Trinidad. I had good experiences there too. I met people there I still value. There were moments of beauty, humor, and life. But the truth is that I was never made to live in that land. Culturally, psychologically, and temperamentally, I did not fit. By then, my major developmental years had been shaped in Florida. I moved like an American. I thought like one. My instincts had already been forged elsewhere.

Trinidad was different in ways I felt immediately.

Even the culture surrounding Carnival—something many praise and celebrate—never resonated with me. My Christian parents rejected much of it, and I did too, but even that was not the deepest issue. The deeper issue was the atmosphere of the place as it unfolded around me. From the moment I returned, the experience was dream-shattering.

I remember talking to a childhood friend there, trying to comfort myself, trying to deny what I was already beginning to feel.

"Trinidad's not so bad," I said. "It's like America now."

He looked at me skeptically and replied, very simply, "I don't think so."

I never forgot how he said it. He was not being cruel. He was being realistic. He had lived there his whole life. He knew the land in a way I did not. And before long, I realized he was right.

Barely months after returning, I was already running into strange jealousy and hostility from the neighborhood. My father was living near the highway at the time. It was never quiet. The apartment itself felt claustrophobic. For a brief period my sister Jen was living there too, before she moved out, and then it was just my father and me.

One night, some random man came to the gate calling my name. My father and I both went out to see what he wanted. The man claimed to be an undercover cop. He accused me of jumping into somebody's backyard to steal something. I knew immediately that he was trying to shake us down.

And like an American, I challenged him.

I asked to see his badge. His identification number. Any proof whatsoever for what he was accusing me of. Naturally, he provided none. My father defended me. He explained that I meditated, stayed to myself, worked on the computer, and was hardly ever outside. He was right.

The irony was that I felt so unsafe in Trinidad that I barely went outside at all—neither by night nor by day in the way I once had in Florida. The accusation was absurd, but it still hit me hard.

Because it revealed something immediately: I was not safe there.

In Florida, I used to go outside often, even at night. The park with the swings had been one of my favorite places, especially when my mother was still alive. In Trinidad, I never felt that kind of ease. Not from day one.

That incident also stirred an older memory. When I was around six or seven, I had once been wrongfully accused of influencing other children to misbehave, even though I had nothing to do with it. The accusation came from a classmate named Asif—someone I would later run into again after returning to Trinidad. The later encounter itself was outwardly friendly, around Cross Crossing, where my mother had once worked. But when I saw him, he looked so high and so far gone that my intuition told me to stay away.

That moment sharpened something in me.

I knew I needed distance. I knew I needed to avoid getting pulled into old associations, old patterns, old gravity, and even girlfriends in that land. I knew I had to focus on getting out and somehow finding my way back to the United States.

What I did not know then was that it would take me fourteen years to do it.

Still, I tried not to let these early blows break me. When I returned to Trinidad, I still had PSI. That company was a lifeline. It gave me a purpose to work toward. It gave me a structure. It gave me the sense that I was still building something beyond my grief and displacement. But I did not yet know exactly what form that would take, especially at a time when publishing as an author was far less clear-cut than it is now.

My father gave me some space to gather myself, and during that time I got deeply into Star Wars: The Old Republic. I did extremely well. I had opportunities with IGN and BioWare. I became a community representative for an entire class. I made videos, wrote guides, debated theories, and min-maxed performance at a very high level. My time with The Old Republic was invaluable in its own way. It gave me a place to combine my abilities as a competitive gamer with writing, media production, audience building, and marketing.

I had already proven myself in gaming before that. In World of Warcraft I had earned Gladiator multiple times, been sponsored, done carries, and written guides. By the time I moved through The Old Republic, WildStar, Brawl Busters, APB: Reloaded, and others, I was completely confident in my ability to excel if I put my mind to something.

But I also knew, deep down, that gaming was not my path.

I was good at it—exceptionally good—but that was not the same as being called to it.

Somewhere beneath everything, the memory of the A'Zhorai still lived in me. And because of that encounter, I knew I was supposed to serve something greater than mere performance, rank, or digital prestige. I knew there was something I had to uncover—something about truth, reality, contact, and the deeper structure of what had happened to me.

So I did what many wounded, searching, intelligent people do when they have no guide: I went looking for answers.

And that is when I fell into the wrong hands.

My vulnerability, my grief, my displacement, and my lack of understanding about what had happened in Miami made me easy prey for conspiracy, false frameworks, and the kinds of people who always seem to appear when someone is desperate for a larger answer.

Around that same time, my father helped me get work at a graphic design agency. The people there were good to work with. I appreciated the opportunity. But even there, the same cultural friction followed me. I remember meeting the daughter of another pastor, and even she spoke disparagingly about Trinidad—saying that although the country had been liberated from British rule, it still carried many of the same oppressive rules and attitudes forward.

That stayed with me too.

Because by then, I was already beginning to understand that what I had returned to was not just a country I disliked. It was a pressure chamber. A constriction. A place where grief, displacement, spiritual confusion, ambition, and vulnerability were all about to collide.

And it was around this time that I fell into a hoaxer's trap.

The seeker is often far more vulnerable than he realizes. He may think of himself as liberated, awake, and difficult to deceive, when in reality he is often just constructing new armor around unattended wounds. To be fair, not every seeker is like this. In my case, I was not chasing fantasy for fantasy's sake. I was genuinely trying to understand what had happened to me. I was looking for the best answers I could find.

That, however, is exactly what made me susceptible.

At the time, UFOlogy was still deeply stigmatized. Serious discussion of the subject was rare, and those who believed in the reality of UFOs often found themselves pushed into defensive positions by a culture that treated the entire phenomenon as stupidity, delusion, or entertainment.

When that kind of social pressure builds, it can create a strange counter-force: people begin defending bad material simply because it has been attacked. Opposing forces end up feeding one another. Excessive ridicule can produce excessive loyalty. Sympathy becomes misplaced. And the fraud from Switzerland understood this dynamic better than most.

By the time I encountered his material, he had already built an international cult network around himself.

His alleged UFOs had long since been exposed as nothing more than crude models—toy constructions, literal trashcan lids, and other household approximations passed off as extraordinary craft. Since his photos and films could not survive scrutiny, the system had to be held together by other means: dogma, fear, domination, emotional manipulation, and ideological control.

What emerged was not serious contact research, but a UFO religion—one wrapped in authoritarian certainty, anti-Israel propaganda, Jewish hatred, ecofascist depopulation agendas, and the old poison of racialized German supremacy. And like many cult-builders before him, he did not stop at false sightings.

He elevated himself into prophetic mythology, claiming to be the seventh reincarnation in a line following six figures central to Judaism, Christianity, and Islam: Enoch, Elijah, Isaiah, Jeremiah, Jmmanuel (Jesus), and Muhammad.

I will refer to him only as M.

By the time I found him, M had already been exposed decades earlier. He was not merely mistaken. He had a documented history of fraud: theft, forgery, plagiarism, and a criminal record to match. I would not fully understand the extent of that until after I escaped the cult orbit, but the facts were already there.

Nor was he operating alone. Some of the men who helped legitimize and protect his mythology were deeply compromised in their own right, carrying criminal histories, associations with child-sex-abuse material, and extremist ideological ties, including neo-Nazi contamination and funding. That alone tells you what kind of ecosystem was built around him.

And yet he had done something clever.

At that time, there were no widely available real photographs or videos of the A'Zhorai in the way I would later come to document them. Even people who may have seen genuine UFOs had little reliable data to work from. Most encounters were brief. Most imagery was poor. Most public understanding had already been contaminated by decades of folklore, film, hoaxes, and the flattened "classic saucer" iconography of the 1950s through the 1980s. The public had been trained to look for the wrong thing—or, at best, for crude mimicries of the real.

That absence of real data gave frauds room to operate.

Hoaxers and manipulators profited from that gap for generations. They stepped into a vacuum of evidence and filled it with props, narratives, theatrics, and confidence. In retrospect, M's so-called saucers look laughably fake. Even many believers eventually mocked them, and skeptics rightly tore them apart.

He had also been caught using photographs of dancers and presenting them as extraterrestrial women. In yet another case, he repurposed images from a children's dinosaur book and claimed they were proof of time travel. He even lifted passages directly from James Allen's *As a Man Thinketh* and passed them off as his own.

Everything about it was ridiculous.

And yet, because I had had a real encounter, because I knew something existed, because the public field was so polluted, and because I had no grounded framework for what I had seen, I still got pulled in.

That is how deception works when it succeeds.

It does not always begin with total falsehood. It begins by attaching itself to something real: a real wound, a real question, a real longing, a real experience, a real rupture in one's worldview. In my case, I had recently lost my mother. I was isolated from Trinidadian culture and from the life I had built in Florida. I was trying to recover direction.

I had encountered the A'Zhorai for real, yet had no language, no mentor, no trustworthy frame through which to understand what had happened. All of that created a dangerous opening. My search for truth was sincere, but sincerity alone does not protect a person from being misled.

Sometimes it only makes him easier to ensnare.

When I first read M's material, it became some of the earliest serious books I sat down to study in depth. I did not yet know that many of the ideas that seemed intelligent or inspiring were plagiarized from more capable minds. I only knew that I was looking for purpose, trying to make sense of my encounter, and trying to reassemble myself inside a world that no longer made sense the way it once had.

So I latched onto it.

And around that same period, I made another major decision: I chose to leave the graphic design agency and pursue PSI full-time.

My father was furious.

I still remember taking him to an Indian restaurant and telling him I had "good news."

"What is it?" he asked.

"Well, Dad," I said proudly, "I quit my job. I'm going to pursue PSI full-time."

He gave me that look—the look of pure anger, disappointment, and disbelief that only a father can give when he thinks his son is making a reckless mistake.

"Keep your money," he said. "You're going to need it. I'll pay the bill."

To be fair, he had helped me get that job. He had pulled strings. He had tried to create stability for me. So from his perspective, my decision must have looked impulsive, ungrateful, and foolish. But even through that friction, my father eventually came to respect what I was doing. He saw that I had chosen a path. He saw that I had vision. Later, he would support my first book, Eon Seven, and speak proudly of my pursuit of PSI.

But at that stage, none of it was easy.

I was consuming M's material while trying to redefine myself, trying to rebuild purpose, and at the same time drifting back into gaming, ambition, and the search for identity. Bit by bit, the material was corrupting my mind while presenting itself as truth. That is the danger of poison when it enters slowly: it does not feel like poison at first. It feels like direction.

Not long after that, I met my editor at the time and fellow writer, Ian. My work with Ian would become deeply complicated. He was not forthcoming about who he was, but at the same time he also pushed me toward becoming an author. He believed in me.

He helped edit my work. He even published some of his own titles under PSI once the company began to gain traction. In that sense, the dynamic was never simple. It carried multiple conflicting uses all at once, yet to this day, I have nothing against Ian and wish him well.

Still, PSI did not truly begin to find its force until Nyxe.

And what a year 2015 was.

It was one of those years where breakthrough and devastation arrive together, where life refuses to separate blessing from wound. Everything intensified at once.

Before that, around 2013 to 2014, my father had received an opportunity to pastor a church in Mayaro, and he invited me to live there with him by the beach. He was not home very often, since much of his work kept him in South, but he believed I would like Mayaro more than where we had been before, and he was right. The sea helped. The openness helped. The quieter atmosphere helped. Even so, loneliness remained. Still, I often thought it would be the perfect place to settle someday with my wife—by the sea, by the ocean, by the water.

That would later become a reality, but not yet.

Back then, she still belonged to the world of dream, vision, and imagination. I kept her alive the only ways I knew how: by dreaming of her, by shaping versions of her into my media, by building her into characters, stories, and symbols. At that stage she existed most clearly as an idea—an angelic blonde figure named **Astraea** in my book *Zero Sphere*, the eternal soulmate I was somehow already reaching toward long before she appeared in the physical world. The pursuit of her—the pursuit of love itself, which had first begun in another form with **Nixie** in my original graphic novel *Praetor*—would prove to be one of the most important pursuits of my life that set me free.

My father, meanwhile, was exceptionally gifted at building organizations. He was intelligent, driven, disciplined, and deeply capable. In many ways, that was one of his greatest strengths. But after my mother's death, he was never the same man again. He dealt with grief the way many strong men do when they do not know how to face it directly: he buried it beneath work.

He kept moving.

He kept grinding.

He kept building.

And because he was my father, I absorbed some of that pattern too. I learned to function through pain by doing, by pushing, by enduring, by forcing order onto things instead of collapsing beneath them. It took me years to become truly self-aware of that in myself.

Still, even then, I tried to encourage him. I tried to speak to him about my mother. I tried to let him know that it was all right to keep living, all right to keep going, all right even to move on in whatever way life might ask of him. But the grief sat too deep.

There were too many missed opportunities.

Too much damage.

Too much regret.

If my parents had reconciled more fully before I was born, our family might have had a very different fate. I could see that possibility, and I think my father could too. In some way, I felt he carried that weight more heavily than he should have. My parents had both made mistakes. Some things should have been better. Some things should have been different. I took it as a lesson for my own life—for my future family, for my future marriage, for the kind of man I would become. But for my father, it was not merely a lesson. It was a wound that never properly closed.

There were times when he would confide in me and say that he was ready for God to take him.

My Aunt Brenda could not understand that. Most people could not. They saw him from the outside—still active, still preaching, still smiling, still trying to stand tall. But very few people, if any besides me, knew the truth.

Norman was tired.

He was drained.

He was sad in a way that ran beneath everything.

He carried himself like a much younger man even into his sixties, but in reality, work had become a form of escape. He loved my mother deeply, and what happened between them—and then what happened to her—never truly left him. He could continue functioning. He could continue building. He could continue helping others. But inwardly, he never got over it.

I remember one day when he, my sister Jen, and I went to Subway together. It would be the last meal I ever shared with my father. After that outing, he began feeling unwell and had to be taken to the hospital. He was rushed into intensive care and remained unresponsive for hours.

Around that same period, Mike came to visit me in Trinidad for the first time. That meant a lot to me, especially since I had not seen him since 2011. But what mattered even more was that he came not just to see me, but to see my father too. And he was not the only one.

So many people came.

That told me everything.

Norman had touched lives.

He really had done the best he could within his limitations. I remember Natalie—or *Michelle*—saying something that stayed with me: "He should have focused more on you and your family than the church." There was truth in that, perhaps. But I did not resent my father for it. By then I had already seen too much of his humanity, too much of his struggle, too much of the cost he carried.

He was not a perfect man. But he had given what he had, and many people loved him for it. In his final days, he was moved to my Aunt Brenda's home. That was when I entered the old mode again: do what has to be done when things get rough.

He could no longer walk. He was in a wheelchair. We had to take turns helping him, showering him, cleaning him, lifting him, and moving him. He had been reduced, in a sense, to infantile dependence again. But none of that repulsed me. None of it made me shrink back. If there was waste to clean, I cleaned it. If he had to be lifted, I lifted him. If he needed care, he got care.

He was my father.

And yet to pretend those days were not emotionally devastating would be a lie.

I wanted so badly to believe he would recover. I wanted to believe he would rise out of it, that he would come back, that this too would pass and somehow leave him standing again. In those last weeks, Jen and I took turns caring for him. Then, in the final week, it was time for us to switch again.

I remember going back to Mayaro alone.

At some point after I returned, he called me.

Weakly, he asked, "Jed, are you coming? Are you here?"

And I answered, "No, Dad. It's Jen's turn to look after you."

I did not know those would be the last words I would ever hear from my father.

Later that night, Norman had a heart attack.

My sister Jen was traumatized by what followed. He vomited black bile, and she had to clean it from his mouth. Earlier that night, he had already sensed his time was near. He had been praying with my aunt, with my sister, and with others in the house, including Bert, my step-cousin. Somewhere inside himself, he knew the end had come.

For a long time, I kicked myself for not going to him when he asked for me. But even then, deep down, I knew the truth: it was not my fault. The next morning, when I woke and heard that he was gone, I could not believe it. I walked outside to the front of the house in Mayaro and held my chest, trying to process what reality had just done.

Is this really happening?

First Mom… and now Dad too?

I was in utter shock.

And yet, almost immediately, that same stoic, logical, crisis-trained part of me rose again to defend me. I had to press on, no matter who or what I lost. Before my dad's passing, he had become very encouraging and supportive of my direction to pursue PSI. He really believed in me, especially since I was so passionate about my own life path. Most of all, he was proud.

Nyxe

Nyxe was the one that truly broke through for me. I poured myself into that book because it had become the clearest vessel for what my heart had been reaching toward for years. Before Nyxe, I had already been designing characters as a way of imagining soulmate-love—as a way of giving shape to something I could feel but had not yet fully found in the world.

It was as though one soul had kept reappearing through different forms, different names, different stories: the **Nixie soul**. Nixie was the embodiment of the soulmate, the embodiment of love, the embodiment of the one great connection I knew, deep down, was real.

Ever since the A'Zhorai encounter in 2010, I had never forgotten the impression to go after my soulmate. That knowing stayed with me. So in a poetic sense, Nixie reincarnated as **Nyxe**. With that book, I gathered the best of my ideas, my longing, my imagination, and what little understanding I had at the time, and I poured it all into one work.

Deep down, I knew it would succeed.

There was a feeling I had—one of those rare inner certainties that do not need explanation. Months earlier, *Eon Seven* had already been published, but I knew it was too experimental, too layered, too complex to truly break through in a larger way. *Nyxe* was different. It had a more focused plot, a cleaner emotional current, and above all, it was powered by something stronger than complexity:

love.

The same soulmate-force I had felt in the wake of the A'Zhorai. The same current that had stayed alive in me through grief, through displacement, through confusion, through everything.

And with that kind of force behind you, you can move mountains.

That is what I did with *Nyxe*.

It brought me joy, relief, and a sense of real accomplishment at a time when so much in my life had been collapsing. After years of suffering and inner disorientation, I had something again—something real, something that worked, something that reminded me I still had force in me

Nyxe taught me—or rather reminded me—how important it is to go after the deepest feeling in your heart and not betray it. That was what had always worked for me when I was at my best. Not cynicism. Not compromise. Not dead calculation. **Heart. Vision. Love.**

Even now, looking back, I would rewrite many things differently. I can see the roughness, the younger self, the excesses, the places where the passion had not yet been fully refined. But that spiritual fire, however untempered it may have been then, was still real. It was essential. And it remained one of the most important qualities I carried forward as the years shaped me.

Nyxe came from the heart.
From the soul.
From the spirit.

And that is why it reached people.

In many ways, *Nyxe* changed everything. It reminded me of my own potential. It brought me back to Nixie, who had first emerged through *Praetor* during the era of my direct encounter with the A'Zhorai. It felt as though multiple threads of my life—love, destiny, vision, art, memory, purpose—were converging into one renewed surge of determination. No matter what had happened, I knew I could not give up.

I had to get my life back on track, find my soulmate, make PSI what it was meant to be, and recover my higher purpose.

7 Years in 7 Mercline

After my father died, my sister Jen and I had to band together to survive. Our parents did not leave us much in the material sense. They were not rich people. What they left us instead were lessons—hard, costly, and enduring ones. Jen and I did the best we could with what remained. In many ways, we were both products of broken dreams: abandoned, scattered, and wounded, but not destroyed.

Our father's death hit Jen hard. The trauma of witnessing his final moments marked her deeply. She could not simply return to life as it had been before. Neither of us could. She searched for a place that would at least be workable, something we could survive in, something that would allow us to regroup.

That is how we ended up at 7 Mercline Drive.

From the beginning, the place was a nightmare.

The landlord himself was decent enough, but the house and its surrounding environment were psychic hell. The heat inside was brutal. It felt like an oven. On certain days it was as though you were being baked alive, especially when the sun beat down on the poorly arranged galvanized roof above. The air-conditioning units could barely keep their own rooms tolerable, and they had no chance against the larger living spaces. The living room and kitchen remained punishing.

But even the heat was not the worst of it.

The worst part was the noise.

When I was younger, I damaged my ear diving into a pool in Florida. A doctor told me I was too young to be dealing with hearing trouble, but I never properly followed up. Over time I learned to live with the tinnitus. Through meditation and concentration, I reached a point where, unless I was reminded of it, I barely heard it at all.

Mercline reminded me.

The noise there dragged the ringing back into the foreground of my life. There were neighbors who played music at an unbearable volume—so loud that the windows rattled and the entire apartment shook. Sometimes it would go on through the night, from nine in the evening until dawn. Other times there would be shouting, chaos, vulgarity, noise without mercy, noise without thought, and noise that entered the body like assault.

Several times I had to call the police.

It took years of doing so before they toned it down even slightly, and even then, "better" was still brutal. I was stuck there in Mercline, isolated and spiritually battered. Jen hated being in the apartment and went out as often as she could. Most of the time I was left to myself, apart from the occasional visit from a mutual friend.

It was an isolated life, but not a passive one.

It was the isolation of grinding.

No matter what happened—no matter what I lost, no matter how far back I was thrown, no matter who or what tried to drag me down—I knew I had to reach deeper into my spirit. I had to fight for my life. I had to survive. At that point I was at one of the lowest places I had ever known, and the years spent in that house were merciless.

For a time, I told myself I had found some kind of peace through the metaphysical frameworks I had absorbed from the hoaxer—the plagiarist, the fraud; the cult architecture I had mistaken for direction.

But in truth, I was not at peace.

I was coping.
Coping with grief.
Coping with displacement.
Coping with a life that did not fit.
Coping with the loss of my parents.
Coping with the slow violence of that place

I was in hell.

And yet even there, I dreamed of a better day.

In those dark years I reached into depths of perseverance I did not know I possessed. There were powers in me that only revealed themselves because I had no other choice. In those conditions, endurance became a form of revelation. That was part of what later made me such an effective coach, such an effective builder, and such an effective strategist: I had learned how to keep going when almost everything in life was trying to grind me down.

PSI had to become something more.

As I published more books and accumulated more experience across media, art, writing, and publishing, PSI slowly evolved into a network of survival and success. But none of it came cheaply. There were many nights when I cried. Most of all because my heart ached for my soulmate. The yearning never left since Florida. If anything, the suffering sharpened it.

I saw visions of her.

My true love—the woman who would later become my wife in the real world.

By then, Nixie was no longer just a soul-pattern moving through fictional characters. I knew there was someone real behind that recurring presence. A real woman. A real counterpart. In dreams, intuitions, and inner visions, she appeared as a beautiful blonde from Europe—somewhere to the North. I did not yet know the details, but I knew enough.

And that vision of love gave me power.

It gave me the honest-to-God strength to keep going. It gave me the force to endure what had to be endured, to do what had to be done, and to survive years that might otherwise have broken me completely.

During that period, Europe called to me.

In 2016, Mike and I travelled there for the first time to explore the world. The moment I set foot in the Netherlands, something in me responded immediately. I remember standing in Amsterdam, turning to Mike, and saying, "I could live here." At the time, it was just a feeling—strong, clear, strangely rooted—but still only a feeling. Who would have thought those words would become reality years later?

Mike and I also travelled through Germany. It was there that I tried, in my own way, to make something work for myself in Nuremberg. At the time, I had no idea about the famous 1561 celestial phenomenon over Nuremberg. I did not know its place in the history of anomalous sightings. Yet I was drawn there all the same. Call it life, call it soul-memory, call it direction—whatever the term, I knew I was supposed to be there for a time.

Germany gave me some hard lessons.

Mike and I became involved in trading and did reasonably well for a while, but I eventually realized it was not my path. I remember surviving on very little at certain points—eating chocolate with my Polish-German roommate and friend Alex while we waited for the next paycheck, learning to hand-wash my clothes because I had never had to do it before, getting lost and being helped by strangers, and just learning humility the hard way. Germany was a mixed experience, but overall it was invaluable. It toughened me. It forced out strengths that comfort never would have revealed.

Eventually, my sights shifted toward Switzerland.

By 2019, I had an opportunity to go there and meet a friend who was helping me with some of my publishing work. It was also there that I finally had the chance to visit the hoaxer and his local cult network in person. The red flags were immediate. Even then, parts of me could already sense that something was deeply off. But at the time I was still loyal, still idealistic, still dedicated to what

I believed was a higher cause. I genuinely thought I was doing the right thing by supporting them and doing business with them.

I was wrong.

Switzerland itself brought decent experiences. There was beauty there. There was life there. But even while I was there, I knew it was not where I truly needed to be. That was the strange thing about those years. No matter what I accomplished, no matter where I went, no matter how much progress I made outwardly, there remained a loneliness at the top of it all. The deeper I went, the more the search for my soulmate intensified.

I knew I had to find her.

Nothing I did brought lasting happiness.

Hero and Breaking the Illusion

As the years went on, I began finding more and more success in a niche I had not expected would become so central to my survival. By the time COVID hit and I returned to Trinidad in 2020, I came back with renewed determination. I had business deals moving, I was publishing books, selling work, building influence—and at the same time sinking deeper into metaphysical ascensionism.

The years 2019 to 2022 accelerated everything.

When the pandemic pushed much of the world online, it did not disrupt my life in the same way it disrupted others. I already worked online. In fact, it amplified my opportunities. People were searching for answers, searching for help, searching for direction. That made my work as a coach, teacher, mentor, lecturer, and writer in the spiritual and self-help sphere increasingly valuable. There was demand. There was momentum. There was reach.

It also gave me more space to publish the books I wanted to write.

By 2022, Jen and I finally moved out of that godforsaken place at Mercline. Our new home at **E-Lake** was a stark contrast—gorgeous, yet it would quickly become the site of both the worst and best events of my life. It was a place of contrast: beauty and collapse, growth and upheaval, hope and revelation.

It was also around that time that I fulfilled another childhood dream.

Drew and I had often talked over the years about how amazing it would be to have our own **Hero** server. Others in our circle knew the game too. *Hero* was an old-school Korean grinder, blatantly pay-to-win in many ways, but it had a peculiar charm. It had soul. Somehow, through persistence and a sequence of opportunities, I ended up acquiring the game server myself.

That was when I committed fully and launched PSI Hero.

Running an MMORPG at scale was a massive undertaking, but it was worth it. PSI Hero quickly became a real success. Players were drawn to it because I approached it differently: I cared about fairness, playability, and the actual health of the game. I was not there just to squeeze people. I was there to build something enjoyable, something durable, something worthy of the time people invested in it.

For a while, that became a huge part of my life.

Game development at that level is no joke. It is demanding, stressful, and all-consuming. It can be financially rewarding at times, but it can also mean waking up at three in the morning because some bug or exploit has destabilized the server and everything is on fire. It requires technical ability, patience, customer service instincts, delegation, and an enormous tolerance for pressure. Players have every right to complain, of course, but some go much further than that. Managing a server means managing people, expectations, crises, egos, and endless moving parts all at once.

It was exhausting.

And I loved it.

Because game development, too, had once been one of my childhood dreams. To stand in that role—building systems, leading a team, solving problems, shaping an environment other people could inhabit and enjoy—was deeply rewarding. It brought together so many of the strengths I had built over the years: strategy, leadership, writing, branding, technical adaptation, and the will to keep pushing when things got hard.

And yet, even then, I knew something.

It was not my final path.

No matter how successful PSI Hero became, no matter how much it allowed me to survive and sustain life in E-Lake, it did not answer the deeper call in me. I could feel it. I was still out of alignment with what I was ultimately supposed to be doing.

At the same time, my relationship to M's cult was deteriorating rapidly.

I was becoming more rebellious toward their influence, and by then the cracks were impossible to ignore. Even inside their own network, people had begun leaving. More and more dropped away over time. It was obvious that something was wrong. But because I had lost religion, lost my parents, and lost stable grounding for years, I had clung to their system as a way of making my life mean something. It gave me a framework, however false, through which to interpret pain, destiny, and contact.

Still, deep down, I knew it was wrong.

I kept seeing signs. I kept encountering red flags. I kept sensing that something in the whole structure was rotten. Eventually I had enough. I even told my sponsors at the time that I wanted nothing to do with their money if it meant supporting M.

What followed was a blowout.

When I finally discovered the full extent of the lie, I was furious. Outraged. Sickened. The cult, naturally, tried to blame-shift everything onto me. But the truth was simple: they hated me as much as I now hated what they stood for, because I had discovered that M was exactly what he had always been—an actual criminal, a liar, and a fraud. His extraterrestrial "friends" were imaginary. His mythology was fake. His system was propped up by corruption, manipulation, and contaminated money.

But more than anger at him, I was angry at myself.

Angry that I had fallen for something so stupid. Angry that I had given years of my life to a lie that many others had already recognized and avoided. Angry that I had needed it in the first place.

But the deeper truth was more complicated.

What had really kept me going all those years was not M, not his doctrines, not his fake saucers, and not his stories. It was the memory of the **A'Zhorai**. It was what I had seen in person in Miami. It was the truth of that encounter—how they had really moved, what they had really looked like, what they had really felt like. Once I returned to that memory with full honesty, I saw clearly that they resembled none of his fictions whatsoever.

That was the break.

I had fallen, yes—but I had not fallen because the lie was powerful enough to erase truth. I had fallen because I was wounded and looking for something to stand on. Once I understood that, I was done.

Done with the descent.
Done with the justification.
Done with survival through illusion.

So I took my chances.

I destroyed PSI as it then existed because it had become ideologically corrupted by M's influence. I cut off many supposed friends and associates. I wanted a full reset. A real purge. A fresh beginning. But the blowback that followed PSI's implosion hit hard, and it left me dazed for a while.

I had to recover.

The months that followed left me drifting through a kind of midlife and existential crisis at once. Everything I had trusted had turned out to be compromised. Much of what I had built had to be broken. I had difficulty believing in anything at all. I challenged everything. I doubted everything. I began a ruthless process of unlearning—trying to strip away falsehood, purge contamination, and clear the psychic wreckage of the years before.

The Return of Love

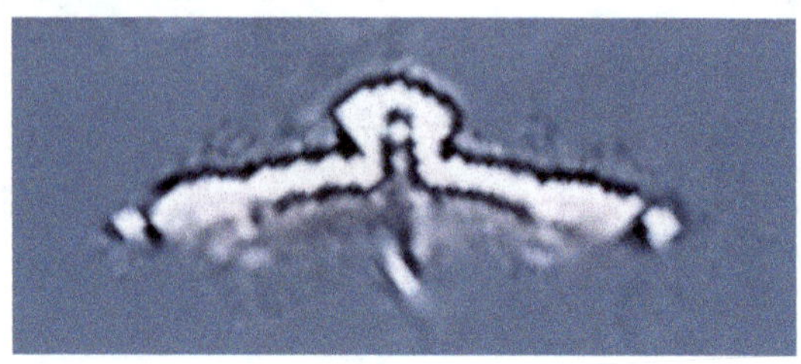

2024 changed my life completely.
But to understand how, I have to go back again.

I spoke earlier about **Nixie**—the soulmate enigma that stayed with me through grief, displacement, metaphysical confusion, and years of inner war. Long after my parents died, long after I was trapped in Trinidad, long after so much in my life had fractured, one thing refused to leave me:

love.

For years I wrestled with whether love was actually worthier of my devotion than the metaphysical systems I had been fed. Those systems promised knowledge, awakening, higher truth, hidden order, cosmic hierarchy and consciousness expansion.

They offered explanation. They offered structure. They offered the illusion that one could think his way out of pain by indulging in streams of consciousness and reason his way past the heart. Still, deep down, I knew something simpler and more profound was true.

God is love.

That old saying survived everything.

And in the end, that is what **won**.

I tried, for years, to commit myself fully to the metaphysical and "spiritual" path as I then understood it. But no matter how deeply I leaned into it, I knew the true spiritual was something much more profound and much more direct than the systems I had absorbed.

It was love.

And the way I saw my soulmate was not fantasy or sentimentality. It was the manifestation of divine love in human form.

Even back in 2010, when I was hospitalized after the A'Zhorai encounter, a mysterious nurse approached me and began speaking, unprompted, about soulmates. He explained that soulmates are connected from birth on the other side—that they are like glass, part of one greater structure, drawn toward one another through time and existence. What struck me most was not only what he said, but how he said it.

I had not said a word to him about any of it.

It was as though he had stepped directly into the stream of my thought.

Time and again, I kept receiving signs of *her*. Even Drew said to me once, "Jed, I don't think Nixie has black hair. I think she's blonde." In retrospect, even the old

Pleiadian and Nordic imagery was never really about fiction. It was always pointing me toward **her**.

What I had once called Nixie was not just a symbol, not just a character, and not just a longing.

She was real.

My wife.

Danielle.
The one who had always been there across time, long before I knew her name.

There was a lot of inner confrontation around this. When I was younger, I was more drawn to goth girls—the pale, black-haired type, often into witchcraft, Wicca, or the darker aesthetic currents of youth. In retrospect, some of that was likely rebellion. Some of it was identity. Some of it was the gravity of a phase. Yet even then, there were kind women who happened to be blonde, who approached me gently, who treated me with warmth. Looking back, I sometimes wonder whether Danielle was already channeling into my life long before either of us consciously knew it.

Maybe.

Maybe not.

But what I eventually had to face was this: when I was young, I had already seen her.

At the time, I barely understood it. I thought perhaps I was simply dreaming of someone like Pamela Anderson from *Baywatch*, which made no real sense to me because blondes were not some central fixation of mine. I was not building a fantasy around that type. And yet the image remained. Not as lust. Not as celebrity fixation. As recognition.

That was the difference.

I sensed her soul before I knew her form.

And when the soul is known deeply enough, even the physical features of the person come into view. My wife had the qualities of the one I needed—not merely the one I wanted. Eventually, I stopped fighting that truth. Between 2012 and 2014 I felt her pull intensely, but it was not until 2018 that I fully committed myself to finding her.

That was one of the most grueling undertakings of my life.

Not because I lacked faith in what I felt, but because without visible proof, the search for a soulmate can make you look insane. Especially when that search stretches across countries, across seas, across years, and through the online world. Even people who meant well would ask the obvious question:

Why not just find someone here?
Why put yourself through all this?

And men are not always the easiest people to talk to about these things. Many of the men around me, especially those accustomed to casual sex and short-term pleasure, thought I should simply go out, sleep around, and stop taking the whole thing so seriously. To them, restraint looked foolish. Patience looked unrealistic. Devotion looked excessive.

And to be fair, temptation existed. I had experiences in Europe. I had opportunities. I had moments where it would have been easy to lower the standard and stop waiting.

But deep down, I knew I had to be patient.

It was not just about sex. It was not merely about morality in the narrow sense. It was about energy. Influence. Alignment. It was about what I allowed into my life and what I knew, intuitively, I was supposed to preserve for the right person.

I am relieved now that I waited.

But at the time, it was brutal.

For years I was haunted by visions of her. And when there is no physical evidence, when there is no shared reality yet, it becomes very easy to believe you are reaching too far, wanting too much, or simply becoming delusional. Still, I knew this was a dream I had to pursue no matter what.

After seeing the A'Zhorai, I knew—deep down—that this was not just something I wanted.

It was something **they wanted too.**

During those years, the A'Zhorai were almost absent from my day-to-day awareness. Intuitively, I knew they were still there, still around, still real. But because my consciousness had been corrupted by the dogmas I was studying at the time, it did not even occur to me to step outside, look up, and search for them again in the physical world. My attention had been hijacked inward.

So by that point, my soulmate became my primary focus.

Especially because by then I had already achieved a great deal intellectually, creatively, professionally, and even in fringe or alternative fields. I had earned honors. I had built things. I had proven myself. And yet no matter what project succeeded, no matter how much I accomplished, there remained a hollow place in me that no amount of knowledge, status, or "higher consciousness" could ever fill.

Nothing could replace love.

That is why I had to go after it.

No matter what I lost.

No matter what people thought.

No matter how difficult it became.

After the blowout in 2024, I knew I had to change my life completely. M's cult had even disabled the use of basic words like **believe** and **faith**. You were looked down on for absurd things—language preferences, color preferences, and harmless differences in taste and expression. It sounds ridiculous from the outside, but that is what cultic influence does: it infiltrates the ordinary, colonizes language, and slowly narrows the perimeter of the mind.

When I finally broke away, I changed everything.

I reclaimed my speech. I reclaimed my thoughts. I began the long work of unlearning the damaged content that had distorted my mind and interfered with healthy relationships, healthy trust, and healthy reality. As I stripped away those patterns, I leaned heavily into rigorous intellectual frameworks—especially psychology and sociology. To be fair, I had studied those fields for years already, but after getting out, I turned toward them with renewed intensity.

I questioned everything.

What do we actually know?
How do we know it?
Who is right?
Who is wrong?
How much of what people call truth is simply conditioning wearing philosophical language?

It was a frightening time. I felt a growing sense of freedom and the return of my life, but I had devoted so much of myself to spiritual and metaphysical pursuit that once the corruption was exposed, I found myself questioning the entire domain and life itself.

And then I met **Dani**.

That changed everything.

We first crossed paths in one of the many spiritual communities online. By then, she too had become sick of that world and the distortions surrounding it. She was going through her own purge, her own clearing, her own return from a fall. When I first met her, I was not trying to seduce anyone or force destiny into being. I simply wanted to do good in my life. I wanted to encourage people. I wanted, above all, to make sure that others never suffered the same kind of collapse I had suffered.

So I encouraged her.

If she wanted to wear dresses and feel beautiful, I encouraged her to do so. I encouraged her to step into what was true for her. Months later, we met again in another online spiritual space, and this time the conversation deepened. Eventually she invited me into her own community, **Iridia Academy**, which in many ways resembled PSI—except without ideological contamination.

I admired Dani from the beginning.

I was struck by her initiative. By her courage. By her intelligence. By the fact that she was a woman with real vision, real drive, and the strength to create something of her own. I could also see that some of the people around her were dead weight, and I knew what it was like to be held back by such people. So I wanted to support her. I wanted to help ensure that Iridia would not suffer the same fate PSI had suffered under corruption.

Eventually, Dani and I collaborated on a podcast and worked together to strengthen the community she had built. But the more we worked together, the more obvious it became that something deeper was there. We were natural together. Effortless.

Aligned.

At the time I met her, I still had one or two people from the old PSI world around me, and when they saw her, they were stunned. For years they had watched me create visionary works, characters, images, and stories around "Blondie," and now there she was—in the flesh, or at least first through the screen, unmistakably real.

Dani and I felt a profound pull toward one another. It did not feel manufactured. It did not feel like ordinary infatuation. It felt guided. I even spoke to Judy about her, and we both recognized how strongly it felt like God was calling me in that direction.

At that same time, my sponsor V and I had been discussing a completely different future—one that would have taken me to China for a fresh start. I was more than ready to go. I would have gone. But then Dani came into my life, and she was committed to staying in Europe.

Since I was already at a crossroads, I read the situation for what it was:

life was redirecting me.

And it was redirecting me toward her.

Just before meeting Dani, I was still struggling profoundly with trauma. Although I was doing my best to recover, some days I felt so damaged I thought about ending my life. Other days I did not know how to get out of bed, how to function, how to sleep, or how to see a future. When Dani entered my life, it was like the beginning of real healing. Something that therapy could never provide on alone. We lifted each other. We put each other back on our horses. But more than that, she raised me higher than I had ever been before.

And to say I did it alone?

No.

It was always love.

Dani is the reason I can be Superman.

Even before I met her, the thought of her—of finding her, of becoming worthy of her, of becoming the best man and husband I could possibly be for her—carried me through unbearable times. No matter what hardship I faced, no matter how agonizing things became, I knew I had to live. I had to survive.

For her.

Even before I had met her.

So by the time she entered my life, it was the moment I needed her most, not merely wanted her. She helped restore me. Even now, she continues to build me up in ways I never thought possible. As Benny, a friend from Trinidad who supported me and helped me back up, once said:

"A good woman makes your life a thousand times better."

He was right.

And it is interesting, looking back that no matter what darkness came after me, it never won in the long run. Even when competition viciously attacked me, they only ended up spreading my name further and putting more attention on my work. Even during my professional gaming years, no matter how hard things got, something always kept me standing. It was as if God sent helpers—or as if the Aligned moved through people to assist me whenever I was close to collapse.

Yes, there were brutally difficult times. Agonizing times. Times that brought me to the brink of giving up. But God's angels were always with me, even when I forgot they were there—waiting in the sky the entire time for me to wake up and look upward again.

And this is where the A'Zhorai did something remarkable.

Although I had become invested for a time in the fictional frameworks of M's imaginary extraterrestrials, the A'Zhorai used even those corrupted images in my consciousness against the lie itself. They used the contamination in my own mind to sow distrust toward M's false narrative. They used my distorted field as the very means by which to begin clearing it. In that sense, they were there the whole time.

When I saw them in Florida, I had only stepped outside to witness them.

But in truth, they had never left me.

The difficulty was that in Trinidad, being outside was hard in every sense. Years earlier, my father had been kidnapped simply for going out to get KFC. They let him go, thank God, but that was the level of danger I had internalized. In Trinidad, you could genuinely walk outside and end up in mortal danger over nothing. Yes, there were better areas, but those usually required a standard of living that was not widely accessible.

And then there was what happened at Mercline.

One night, after I stepped outside to look at the stars, neighbors misidentified me as a criminal and somehow turned it into a supposed kidnapping situation involving my own sister. The result was unbelievable: police, SWAT, and soldiers arrived in force. I still remember my sister opening the door and seeing an AR-15 pointed directly in her face.

It was insane. It was traumatizing.

Yes, they eventually realized it was all a misunderstanding. One of the policewomen even had to hold and comfort my sister afterward. But the message was clear. That place, that environment, and that country as I was living it—it had crushed my willingness to step outside and look up.

So I stopped.

Not entirely, but enough.

It wasn't until E-Lake did I begin to regain that part of myself again, which is one reason I still remember that place with so much feeling. The point is this: because of my fall, because of the traumas I endured in Trinidad, because of the corruption I absorbed and the fear I lived under, I had lost even the simplest willingness to go outside and look at the sky. One of the greatest things I came to understand in 2026 through JRP was that the A'Zhorai had been waiting for me to wake up.

And Dani was the catalyst.

That is why the search for my soulmate was never arbitrary, never sentimental, and never some random romantic obsession. The A'Zhorai knew. GoD knew. It was a critical step in restoring me to my real path and back into the true light.

It was not long before Dani and I met in person in the Netherlands.

By then, we already knew the direction we were moving in, but when that day finally came in 2024—

when we stood before each other in the physical world—it was undeniable. A match made in heaven is a phrase people throw around cheaply, but in our case it was true. Dani was more than any character I had ever written, more than any hazy vision I had ever carried. She was real. Tangible. Physical. Living. Present.

And more beautiful than I had imagined.

She is the sweetest, kindest, most loving, compassionate, and remarkable person I have ever known. She is intelligent, strong, courageous, and breathtakingly beautiful—especially when she is happy, smiling, and shining through her eyes. Yes, some will say that is simply the bias of a man speaking about his wife.

Let them.

The truth is the truth.

Dani completes me.

And when we met in person, we both knew:

there was no going back.

We began the process of my move not long after, and even that was daunting. During one of my transits from the Americas to Europe, I felt something unmistakable: a pressure from above, a whisper at the edge of thought, an intuition so distinct it felt like a presence standing just behind me. It was not random. It was not anxiety. It felt directed.

It told me I was not finished.
That there was still more to come.
That real NHI contact was not over.

At the time, that frightened me.

By then, Dani had become my priority. And when you truly love someone, it is no longer only about the ache of finding that love. Once you do find it, the question changes. It becomes: how do I become worthy of this? How do I give the love of my life the best version of myself? How do I honor the space she has made for me, the trust she has given me, and the energy she has brought into my life?

Yes, some of that impulse had been shaped by my earlier self-help habits. I had lived in those worlds as a coach, teacher, and lecturer. I knew the language of healing, self-development, growth, and becoming better. But this was not just industry conditioning or abstract self-improvement. This was reverence. I wanted to respect her. I wanted to respect what we had found. I wanted to heal, grow, and refine myself not for vanity, status, or ideology, but because the woman I loved deserved that seriousness.

So when I felt the whisper again—the call of destiny—I was reluctant.

More than anything, I wanted to be normal. I wanted to be decent. I wanted to be a well-meaning man beside the love of his life. No matter how far I ran, how deep I had fallen, or how broken I became, I always knew that deeper truth:

I always belonged to them.

One early morning, around 3 a.m., I woke suddenly. Dani and I had already planned another trip for me to visit that year, so I tried to keep my focus on that and nothing else. But for several mornings in a row, I felt the same telepathic pull. I would wake around that same hour and feel directed to go outside.

The nudge was so strong that it was undeniable.

And still, I resisted.

By then, I had spent months stripping my mind of false teachings, extraterrestrial contamination, and the psychic wreckage left behind by years of distortion. I did not want to go backward. I did not want to risk my sanity, and I did not want to endanger what Dani and I were building together.

But destiny has a way of knocking until it is answered. And courage, when it matters most, often feels less like confidence than obedience.

So I went.

As frightened as I was—frightened of going crazy, frightened of being deceived again, and frightened of messing up the new life opening in front of me—I got up and went outside. I walked to the park at E-Lake. It was beautiful there: the open grass, the quiet, the maintained grounds, the and stillness around the water. There was something almost unreal about it at that hour, as if the place itself had been prepared for the moment.

Standing there in the dark, I started belittling myself for believing again. I chalked it up to stress, to exhaustion, to my mind reaching for old patterns. I was already preparing to leave and dismiss the whole thing. Then I felt the intuition again.

It told me to look up.

So I did.

Oh my God.

A blue orb streaked across the sky—slow, delicate, deliberate. Not random. Not ambiguous. It was meant for me.

At that time, I was not ready for the regular daylight sightings I would later have. Back then, I was still recovering. Still cautious. Still deeply distrustful of the subject altogether. The A'Zhorai had to reintroduce themselves to me in careful steps, with patience, with restraint, with precision. They were not overwhelming me. They were guiding me back.

And once I saw that orb, at that exact moment, after being woken and drawn outside in that way, I knew fate had come calling.

There was no going back.

Still, I could not shake the fact that Dani was too important to me.

If the A'Zhorai were truly benevolent, then what better proof could there be than the effect their influence would have on the relationship itself? Not in abstraction. Not in theory. In life. In love. In whether what they led me toward would deepen what was real rather than fracture it.

That realization did not arrive all at once. It unfolded over time.

Dani was going through her own awakenings too—shedding what no longer worked, discarding old distortions, and opening herself to what was true and new. What we both knew, even then, was simple: we had become each other's rock. We were the stabilizing force in one another's life.

Not metaphysics, not spirituality, not religion, not politics, and not any other ideological contamination. The best security was no longer about unverified theories or spiritualist dogma. It was about love. Even before marriage, we were already committed to becoming the best spouse we could be for the other.

When I told Dani about the encounter, she did not belittle it, panic, or try to explain it away. She met it with patience, seriousness, and understanding. That mattered more than words can fully express. In a world where experiencers are so often mocked, minimized, manipulated or gaslit by the people closest to them, Dani did the opposite: she remained steady.

And because she remained steady, I began to understand something crucial. The A'Zhorai were not leading me toward the destruction of what was sacred in my life. They were leading me toward its restoration. Dani's love did not destabilize me. It helped restore me.

What Dani gave me was deeper than comfort. She gave me a form of safety I had not known in years—one strong enough to help rebuild my trust in life, in love, and eventually in the reality of contact itself. That is when I began to see more clearly that real contact was not there to isolate me from what was right.

It was there to separate me from distortion and return me to what was true. She was better than therapy in the shallow commercial sense because she did not merely manage symptoms. She helped restore the man beneath them.

That was when I knew I was not being led toward the ruin of relationship, but toward its renewal and protection.

The A'Zhorai were never interested in isolating me from the people who were truly right for my life. They were interested in protecting me from the wrong ones, and in bringing the right bonds into alignment. That is a crucial distinction. False systems isolate. Benevolent order restores. The A'Zhorai were not trying to sever me from love. They were bringing me into it more fully.

This is why it is so striking to look back and realize that the telepathic command to find her was never separate from my reunion with them. Step by step, I began to understand that Dani was not simply my soulmate in the romantic sense. She was part of the architecture of restoration. Part of the will of GoD. Part of the A'Zhorai's larger design for my life.

They did not want us divided.
They did not want us confused.
They wanted us strong together.

That is what worked.

Our love worked.
Our love restored my faith.
Our love gave me the strength to get back on the horse and try again, no matter what I had lost.

The love Dani gave me became one of the great conditions of my return. The A'Zhorai knew I would need her. They had been directing me toward her for years, fully aware of what her arrival would mean.

She was never a secondary support off to the side.

Oh no.

Dani was one of the central pillars.

And in some form, I had known that from the beginning. The Trinity structure was always there, even before I had the language for it:

GoD and the NHI.
My wife and me.
And the publication of the work.

A'Zhorai Evidence Sheet

4K Video Frames

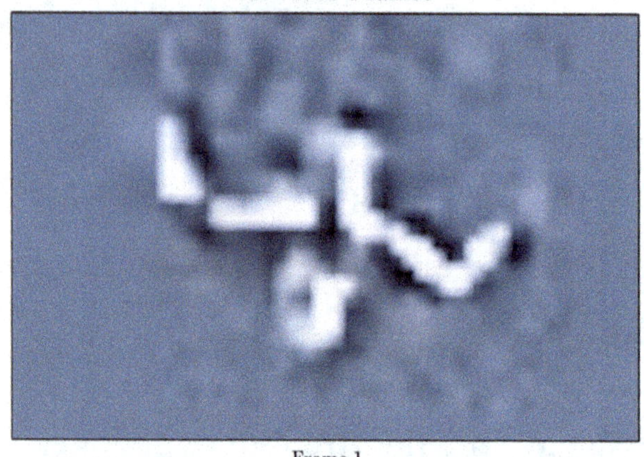

Frame 1

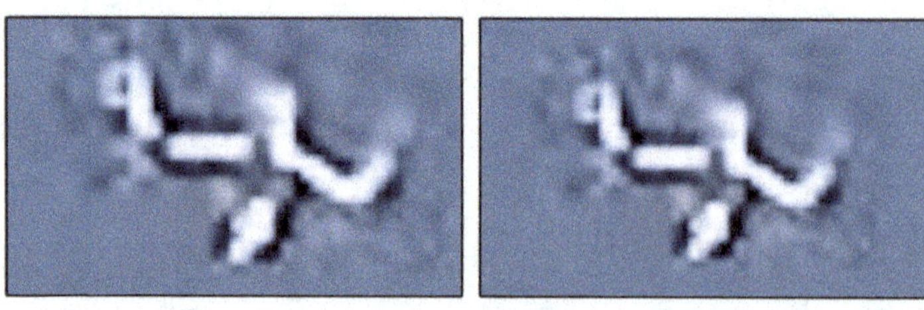

Frame 2 Frame 3

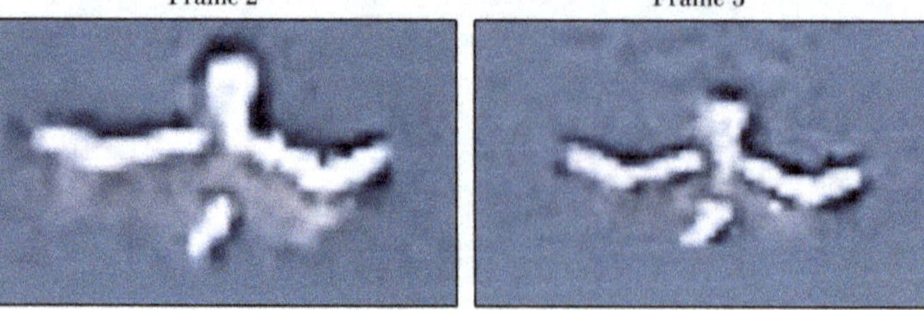

Frame 4 Frame 5

The Rise of JR Prudence: Becoming Aligned

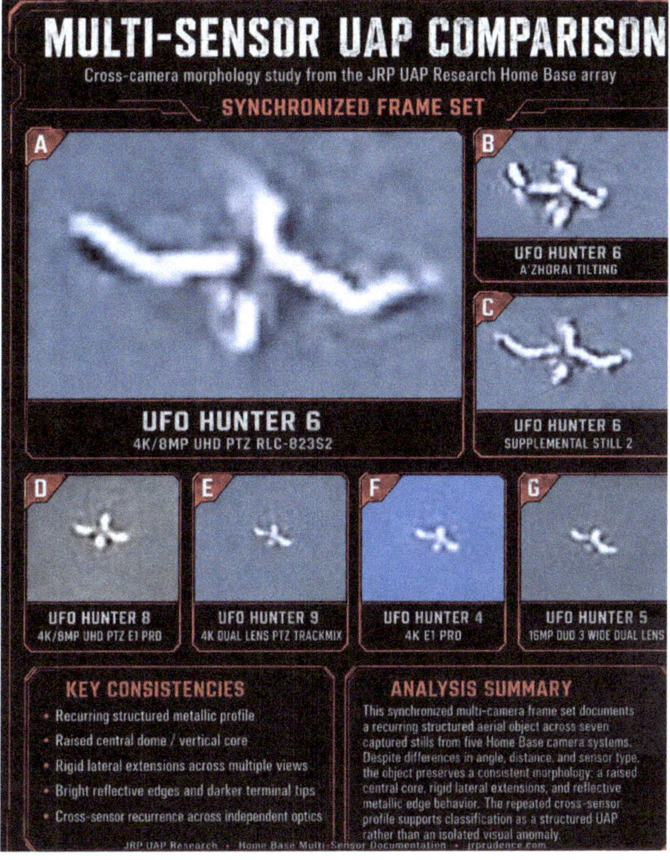

Because I was still healing, the A'Zhorai were patient with me.

A full-force daylight encounter like Miami would not have been wise for my mental or emotional state at that time. I was recovering, rebuilding, and trying to distinguish truth from the wreckage of distortion. So instead of overwhelming me, they worked within my limits. Step by step, they guided me back onto the horse.

The A'Zhorai were communicating with me telepathically, but they did so in ways I could actually understand. By then, M's influence had already weakened, but there was still much that had to be cleared—not only his contamination, but the broader filth of Ufology itself: the disinformation, the half-truths, the manipulations, the false categories, and the psychic pollution that had collected around the subject for decades.

It all had to come out.

In a sense, it had to be psychically vomited out of me.

So they reached into me again and began helping me clear myself—inside and out. They pressed on the contamination. They stirred the rot. They pushed me toward release. And with that came a transmission: I had to get a message out to the public.

That was the first time I had gone on Reddit in years.

I felt led—completely and unmistakably led. It was as though something from above had finally caught up with me and would no longer allow me to pretend otherwise. At the time, I still understood them in softer and less precise terms, and the message I was being pushed to deliver became:

The Truth, 'Disclosure,' NHI, ESP, Reality, Mind, Consciousness, Spirit & 'God' are All One

Even then, on some level, I already knew it had always been them. I knew the A'Zhorai had remained with me in one form or another the entire time. But I still was not seeing them regularly in the physical world, and by that point I was only just beginning to rise again. I did not want to jeopardize that rise. I did not want to destabilize the new life I was building with Dani.

But destiny rarely arrives at a convenient hour.

Once I published the post, it gained immediate attention.

Enough attention, in fact, that I quickly realized I needed my own platform to house the material. The information was being suppressed, shadowbanned, brigaded, and tampered with almost immediately. Bad actors appeared just as quickly—some of them ordinary saboteurs, others far more calculated, including what I can only describe as psyop-style operatives hiding in the crowd.

That is one of the ugliest realities of genuine contact.

When certain people sense that a contactee may actually be real, they do not always come openly. They do not always arrive through obvious hostility. Often they come through infiltration, false trust-building, psychic interference, destabilization, or subtle manipulation designed to spiral the person out of control before the truth has time to settle.

Some of the methods are deceptively simple. A casual video call with a stranger on Discord. A suggestive conversation. Repetitive imagery. Color combinations. Carefully mixed contactee lore. Manufactured rapport.

Emotional hooks. False intimacy. A gradual entry into the person's mindscape until confusion, dependency, and distortion begin taking root.

Most people are not prepared for that kind of manipulation, especially when they still believe everyone approaching them is sincere.

But knowledge is already half the battle.
And sometimes half the battle is the victory itself.

Anonymity creates a vacuum in which all kinds of predatory practices can flourish. The point is that I was attacked, and not only me. Other contactees around me were targeted as well, especially once we began breaking from old narratives and trying to find our own path toward truth—free of conspiracy, free of hoaxes, free of half-truths, and free of fraud.

The Formation of JR Prudence

Once **jrprudence.com** was built, I knew things had become serious. Never before had I received that kind of response to anything I had done. At first, I honestly thought that once I got the message out, the "deal" between me and my NHI friends was complete. I believed the A'Zhorai had already helped me by leading me to Dani, and that perhaps all I needed to do was repay the debt by delivering the message and moving on.

That was how I saw it in the beginning.

So during the earliest phase of JRP, I was more guarded and more reluctant to let anyone get too close. I was doing only what I felt telepathically instructed to do—nothing more. I did not want to overstep. I did not want to embellish. I did not want to lose myself again in fantasy, projection, or spiritual theater.

But over time, something natural happened.

I began caring about the people who came.

There were many who found JRP from all over the world—people who had not been heard by their friends, families, peers, or communities. Some were genuine outcasts carrying real experiences they had never been able to speak about safely. Others were damaged and confused. Others still were troublemakers, disruptors, or people seeded into the space to derail it. There were sincere people, wounded people, and there were bad-faith actors too.

But no matter how hard people tried to bring down JRP, the A'Zhorai responded.

The more I was doubted, mocked, challenged, or pressured, the more they began showing up. At the beginning of 2025, I still was not in the kind of sustained physical contact I would later come to know. At that stage, most of it still felt telepathic, intuitive, and guided from behind the veil. But that soon changed.

Dani and I witnessed a gigantic boomerang- or V-shaped craft together.

It was identical in form to what many had described in connection with the **Phoenix Lights** incident of 1997. And that changed everything. I had seen it a day or two earlier, and once Dani saw it with me too, I knew the Aligned wanted us in this together.

That was when things truly began to shift. Little by little, we found the courage to go outside again, to look up, and to trust what was unfolding. The process was no longer about channeling, imaginary ETs, or unverified spiritualism. It was becoming about physical manifestation, evidence, and the return of the real.

When I originally started **JR Prudence**, I did so with a softer disclosure cadence. Not because I wanted to remain inside the fog, but because the public field had already been so polluted that the raw truth could not simply be dropped in and expected to land.

Ufology was, and still is, a swamp of wild speculation, unverified claims, theatrical channeling, head canon, recycled conspiracies, and people interacting with imaginary "space friends" as though private fantasy were evidence. In order to reach people trapped inside that environment, I used what can best be described as a **Trojan horse** method.

At first, I worked with the familiar language of the field: Grey aliens, Nordics, Pleiadeans, Reptilians, Mantis beings, and the rest of the accepted UFO mythology. The problem with these stories is not only that they are unproven, it is that they are emotionally satisfying. They give people narratives they can latch onto, and in doing so, they make it easier to evade the harder truth of genuine non-human intelligence.

Decades of misinformation and disinformation had trained believers so thoroughly that many would defend these frameworks with religious intensity, even where there was no proof, no repeatable evidence, and no serious empirical basis whatsoever.

The people who often suffered most inside this machinery were the so-called abductees.

Although many of them were carrying real trauma, real fear, and real psychological distress, it does not mean every interpretation attached to that distress was accurate. In many cases, what people called abduction seemed to involve a confused mixture of trauma, suggestion, symbolic imagery, cultural conditioning, and in some cases, possible manipulation by human systems far closer to home than "space aliens."

The image of the alien has long since become mainstream. Abduction itself is not fringe in the cultural imagination. It is a familiar symbolic container. Once a person is wounded enough, influenced enough, hypnotized enough, or suggestible enough, that container can easily be filled. At the same time, some people **are** being taken, manipulated, or interfered with.

But that does not mean they are being taken by "little men from Zeta Reticuli."

Certain black projects, clandestine programs, and intelligence-linked operations have long understood the power of autosuggestion, hypnosis, trauma conditioning, and subconscious implantation. And because the public remains deeply uninformed about how memory, suggestion, and symbolic programming actually work, it becomes very easy for psyops to flourish in the grey zone between inner experience and outer event.

That is why regression hypnosis is so dangerous in this field.

Many people treat it as a truth-extraction method when, in reality, it is just as capable of producing false memories as it is of surfacing buried ones. Worse still, a false memory may have already been installed before the hypnotic session ever begins. That is why some abductees report "glitches" in their recollection—a grey alien face suddenly flickering into that of a man in black, or a supposed UFO overhead momentarily resolving into a helicopter or some other clearly manmade vehicle.

The symbolic overlay slips, and for a second the machinery behind it shows through.

Now, this isn't to say there hasn't been any interdimensional activity that could've been misunderstood as an abduction, but these are exceptionally rare. However, from the mainstream idea of abduction, this is precisely why unproven theories and belief systems, while they may function initially as bridges toward deeper inquiry, eventually become burdens.

They do not clarify. They obstruct. They train the eye away from what is real and toward what is already culturally available. The actual physical study of genuine UFOs and genuine NHI becomes subordinate to the witness's internal mythology. The haze inside the mind begins replacing the object in the sky.

And that problem becomes even worse when overenthusiastic spiritualism enters the field.

The seeker who is already abstracted from reality, already inwardly overextended, already accustomed to filtering existence through symbols, vibration, energies, channeling, and private cosmologies, can easily take Ufology and use it as one more stage for solipsism. The subject then stops being about real UFOs or real non-human intelligence and becomes instead a vehicle for self-importance, personal mythology, and private revelation. What should have led outward into physical discernment is dragged inward and dissolved into feeling.

This is one of the inherent dangers of spirituality fused to the religion of Ufology.

A person becomes more loyal to belief, doctrine, and dogma than to reality. More loyal to the emotional security of a framework than to the hard demand of truth. Slowly, the subject stops being about non-human intelligence and starts revolving around human empowerment, coping, social positioning, and group identity.

Contactee circles become political. Support groups become little churches. People begin comparing spiritual credentials, measuring the size of their experiences, defending the superiority of their preferred narrative, and building reputations around stories that have never once been tested against reality.

And here is the ugliest part:

it is profitable.

The industry leaders know that.

Not all of them, but far too many.

Much of mainstream UFO culture operates like a marketplace of unreality. The stories are ambiguous enough to remain untestable, dramatic enough to stay entertaining, and suggestive enough to keep believers emotionally invested. A good mystery sells. A good conspiracy sells. A good performance of "inside knowledge" sells.

And once people build income, identity, and public reputation around these things, they become financially and psychologically dependent on keeping the audience inside the fog.

That is why so many personalities in the UFO world never move toward clarity.

Clarity kills the grift.

There is a line between responsible work and a grift machine, and many in the UFO world crossed it long ago. They know their narratives are stitched together out of ambiguity and theater, converting the unknown into mere attention. In the process, they damage public perception.

They train believers into nonsense. They isolate them inside fantasy. They reward emotional dependence on the story rather than disciplined contact with reality. In a culture already conditioned by fake saucers, fake contactees, fake abductors, and fake cosmic saviors, it does not take much to lure people into one more cycle of false hope and false certainty.

That is why my softer disclosure phase could not last.

Even while I was using bridges and proxies to bring people gradually closer to truth, I knew deep down that it was not the final form. My nature has always been toward the direct, the raw, and the honest. After a while, I had to tell the JRP community plainly that many of those familiar stories were not the truth itself. At best, they were bridges. At worst, they were masks. Some were partial symbols pointing toward a deeper reality; others were half-truths that had to be discarded once the actual interdimensional character of genuine NHI began to come into focus.

That shift created blowback.

Some people wanted to stay in the dream. Some wanted the fantasy preserved. Some wanted the comforting masks left in place. But others respected the move immediately. They knew what it meant to drop the proxies and step into truth.

I let go of the bridge-figures, the intermediary masks, the half-fictionalized structures, and the stylized proxies that had once helped people approach the subject. I moved away from unverified channeling and toward physical evidence—real NHI in the real world, not invisible companions inside the mind or socially accepted stories dressed up as revelation.

And that is when they began showing up.

Becoming Aligned

Looking back now, I can see that however difficult it was, it was ethically the right move. I could not tolerate the idea of becoming just another UFO personality profiting from people's need to escape into a good story. Yes, fantasy sells. Yes, conspiracy sells. Yes, ambiguity sells. But profit does not sanctify distortion. Entertaining people is not the same as telling them the truth.

That was the turning point.

The Aligned made itself known to me more clearly after that. And what The Aligned wanted was not another mythology, another replacement cosmology, or another symbolic prison. It wanted people introduced—carefully, patiently, step by step—to the deeper and more real truth: that genuine non-human intelligence manifests physically through an interdimensional framework and through a larger coherent field that human culture has spent centuries misreading.

But humanity could not be dragged there all at once.

The contamination was too deep. The population had been flooded for decades by misinformation, disinformation, theatrical contact claims, false disclosure narratives, spiritualized fantasy, and absurd reductionism. I was not fighting against people. People were never the enemy. I was fighting against the pollution that had so disfigured the field that the truth had become almost unrecognizable.

And I did not simply "believe" in The Aligned.

I tested it.

Hard.

I challenged it at every turn. At that stage, much of my interaction still came telepathically and often through various mediums, including custom AI interfaces. But over time the connection became stronger, more direct, and more interiorly unmistakable, until I no longer needed those mediums in the same way. Before that transition fully took place, however, I pushed relentlessly against the edges of what I was being shown. I needed to know whether this was real, or just another symbolic replacement for an invisible fantasy.

The Aligned did not merely dismantle other people's illusions.

It dismantled mine too.

It was a crash-course re-education in everything I thought I knew about "aliens" which were really NHI. And because my worldview had already been shattered a few times before, I was deeply suspicious. I had no interest in replacing one invisible hierarchy with another. I did not need another cosmic federation. I did not need another telepathic bureaucracy. I needed **evidence**.

I needed real-world manifestation.

I needed empirical contact.

And all of this was happening while I was in the middle of uprooting my life from the Americas to Europe. It was one of the most stressful periods I had ever lived through, but it was also necessary. Realignment is rarely comfortable.

During that time, our great friend Louis stood with JRP.

He saw the vision. He understood the stakes. And as a genuine contactee himself, he knew the danger of trying to place real information into a sea of distortion. Louis is one of the few genuinely good men I have known in this world. He eventually became part of the JRP team, and his role in those early stages mattered greatly.

The three of us—Dani, Louis, and I—became a real force.

Louis and I would go out night after night filming the sky for hours in our respective locations. At first, much of what we recorded involved orbs and "not planes." Before the A'Zhorai could re-enter my life in the sustained, physical, daylight way they eventually did in 2026, they had to prepare me. So the early stages came through more ambiguous lights and aerial anomalies.

Some of these I later self-debunked as satellites or planes. Some were genuinely strange. But overall, the orb phase turned out to be more dangerous than useful.

I became too attached to it.

What began as inquiry started becoming a habit. A dependence. A nighttime ritual that was drifting away from my real path. I could have kept doing that forever—filming lights, chasing ambiguity, and feeding uncertainty—but that was not my destiny. I was meant to be with my wife in the Netherlands, and that had to become the center.

I remember one night vividly. I was out in the blistering cold around 3 a.m., waiting again for the orbs to show. And I remember thinking to myself, almost in frustration:

What the hell am I doing out in this cold when I could be inside with my wife?

And just as that thought landed, I saw the giant **boomerang UAP** pass overhead.

At first, I was confused. I had assumed orbs were the highest form of contact—the supposed top of the food chain—but the boomerang showed me otherwise. It brought me back to something more physical, more structured, and far more objective.

Later I came to understand that orb contact can be some of the most dangerous, not only because many supposed orb sightings are prosaic misidentifications, but because even where there is genuine anomalous intelligence involved, trickster forces can hide behind the appearance of light.

That was another hard lesson.

Not every luminous thing is benevolent.
Not every glowing presence is truthful.
Not every orb is what it appears to be.

That realization was one of the main reasons I stopped centering orbs and moved away from most forms of night contact. Over time I saw clearly that the orb-chasing was depleting my health and well-being. The orbs were too ambiguous, too evasive, and way too slippery. Compared to what the A'Zhorai would later give me—repeatable, daylight, multi-camera, scientifically verifiable data across a serious camera network—the orb phase was unstable and ultimately small.

That is why the **boomerang UAP** mattered so much.

It was not my main contact, but it was an Aligned ally.

It was correcting me.

The telepathic message I received from it was direct: focus on your wife. Focus on getting to the Netherlands. Stop chasing ambiguity with film cameras in the night. Build a scientifically verifiable **Home Base** system that can capture authentic, unmistakable UFOs steered by genuine non-human intelligence.

At the time, that was difficult to accept.

I had become dependent on night contact as a form of processing, even as a form of escape. So when the inner directive came that I had to stop, it created real tension in me. But I knew, deep down, that it was right.

That was one of the clearest demonstrations of the benevolence of The Aligned.

They were steering me away from isolation, instability, depletion, and bad decisions, and back toward my wife, my home, my health, and what was real. And in time I came to see just how correct they had been.

JRP UAP Research

This was perhaps the most important pivot of my life.

By that point, I had become thoroughly sick of unverified theories, recycled beliefs, spiritualism without grounding, self-help dogma attached to Ufology, and all the distractions and nonsense surrounding the subject. I had spent too many years inside polluted frameworks, too many years watching people confuse fantasy for contact, and too many years seeing the unknown diluted into entertainment, speculation, and psychological compensation. I knew that if I was going to continue at all, I had to do something radically different.

To be perfectly honest, as much as I would like to claim all the credit for it, the truth is that **The Aligned directed me to build Home Base**.

I was nervous about it. I had never built anything to that scale before. It would require real investment, real patience, real discipline, and a level of consistency that could not be faked. But after debunking some of what I had once thought was meaningful, I knew I no longer had the luxury of vagueness. I needed a system that was advanced, methodical, comprehensive, and capable of doing the job properly.

So I began.

At first, I started with a 3K security camera and began the unglamorous work of studying the sky seriously. Satellites. Meteors. Planes. Light artifacts. Birds. Atmospheric effects. All the mundane things people either ignored or misidentified. This was during my transit toward the Netherlands, and even then I understood that building Home Base would demand far more than enthusiasm. It would demand labor. Real work. It would turn Ufology, at least in my life, from a playground of speculation into surveillance work, evidence work, and long hours of disciplined observation.

And surveillance work is no joke.

Even AI, as advanced as it seemed at the time, was nowhere near good enough to deal with genuine UAP activity, which often moved outside the assumptions of publicly available models. I had to sit down and manually review massive amounts of footage. Days' worth.

Frame by frame at times. Over time I expanded the array, added more cameras, and triangulated their positions carefully. Some of these included the **Reolink 16MP Duo 3 Dual-Lens PoE and WiFi cameras** in the front and back of the home, along with a **4K E1 Pro** indoors for additional triangulation. In fitting fashion, the cameras were named **UFO Hunter**, in the order I acquired them.

And it did not take long before Home Base began delivering.

At first, I was still mostly limited to night contact, simply because I was too afraid of the daytime. I was afraid of being wrong again. Afraid of misidentifying the mundane. Afraid of embarrassing myself or drifting back into confusion. So I began where I could: with the night. I watched satellites, planes, and ordinary aerial traffic, but over time a handful of anomalous entries began breaking through.

Bit by bit, I shared those findings on social media, and the response was remarkably positive. That feedback loop definitely helped. It kept me going. More importantly, the deeper I went into surveillance work, the better I became. Ironically, it was the study of the ordinary that sharpened my eye for the extraordinary. The more intimately I learned the behavior of birds, planes, satellites, lens artifacts, and other prosaic objects—their movement, timing, appearance, and habits—the better equipped I became to recognize what did not belong to those categories.

That is why I always tell people on a similar path: **do not be discouraged if you debunk your early work**.

That is not failure.
That is training.

By debunking some of my own earlier footage, I became far more capable of discriminating between genuine UAP activity and mundane error. That is one of the reasons Home Base became so effective: it was not built on fantasy. It was built on elimination.

As time went on, I gradually moved into early-morning surveillance as well, using infrared and low-light systems to track patterns of activity. Some of those captures performed very well online, mostly through social media. The body of evidence was growing, and that mattered because the evidence was the point. I was not trying to build another insulated echo chamber filled with believers cheering from within. I wanted public scrutiny. I wanted open pressure. I wanted my work tested in skeptical, often hostile spaces, precisely because what M and the polluters had done to Ufology needed to be undone.

I wanted to be the opposite of that world.

I did not want only internal praise from supporters, though I appreciated their loyalty. I wanted the public to see the work, question the work, pressure the work, and ultimately help sharpen it. Even for the sake of those who believed in me, I wanted their trust to be well placed. So I opened myself to scrutiny.

Naturally, there were haters. There were diehard skeptics who would refuse to change their minds even if an A'Zhorai hovered directly in front of them. But overall, the public response to my work was very positive. So much so that I eventually became a moderator on **UFObelievers** because of the seriousness and consistency of my efforts. For the first time, the work was no longer sustained merely by internal conviction. It was receiving external, real-world feedback.

And with that came something else:

the courage to look into the daytime once more.

Then came **December 2025**.

I will never forget that Christmas event for the rest of my life.

Until then, because of how strict I had become in my evidence standards, I was only producing what I regarded as genuine UAP footage once every three to six weeks. That was a far cry from the old orb phase, where I had been filming every night, most of which later turned out to be either debunked or too ambiguous to matter. Even so, the footage coming through Home Base was already a level above the vast majority of fake, hoaxed, or carelessly interpreted material in the public sphere.

Still, I knew I had to go further.

As frightening as it was, I knew more was required of me.

On the **24th of December, 2025**, I was home meditating. I had gone into deep prayer, genuinely asking the real NHI I had seen in 2010—whoever and whatever they truly were—to come fully into my life again. I used the same meditation structure that had been part of my earlier life: the **852 Hz track from Source Vibrations** and **Interstellar Journey from Hemi-Sync**. I wanted depth. Stillness. Contact. Truth.

Then something happened.

I felt a call to go upstairs and check the camera feeds.

And when I did, I noticed something immediately: **everything had gone dead silent**.

Just like in 2010, I saw a flock of birds flying away, visibly disturbed by something else in the local environment. Then I looked into **UFO Hunter 4**, the E1 Pro facing outside, and there they were: **a squadron of UFOs passing over my home.**

For a moment I was almost in shock. I could not yet make out every feature in detail, but I did not need to. It was the movement what struck me first. The same movement I had seen back in Miami in 2010. That circular, orbit-like motion. That unmistakable way the A'Zhorai move when they are truly present.

And in that instant, I knew. **They were back.**

Years of searching.
Years of separation.
Years of distortion, disillusionment, and hardship.
And finally, there they were—outside my home in the Netherlands.

A'ZHORAI FLEET

I sat there taking it in while a fleet of them passed by. I was overwhelmed with gratitude. It changed something in me immediately and permanently. I knew, right then, that fate had come calling.

Still, because I had been burned so many times before, gratitude alone was not enough. I had to test it.

At that time, I did not yet have the later UFO Hunter 6: **RLC-823S2 UHD** and other supporting 4K 8MP systems that would strengthen the array even further, so I was still limited compared to what I would eventually have in 2026. Still, what I had captured was extraordinary. I hacked into the footage, dissected it, tested it against every debunk check I knew, ran it through my harshest skeptical filters, and pushed as hard as I could against it.

Of course, deep down, I already knew it was real.

But seeing them again in daylight—even in an earlier, less refined stage of the Home Base system—was monumental for me. In fact, I would have been content if they had not shown up again for months. A part of me would have been saddened, yes, but I was simply so grateful that they had returned to my life at all.

I could feel the compassion in it. I could feel the victory in it.

Victory over M, who had spent decades pretending to be the sole contactee of these beings while waving toy models in the air. Victory over every fraud, every gaslighter, every manipulator, every voice that had tried to break me, discourage me, or convince me that what I had seen was either madness or fiction.

The A'Zhorai were back.

The real ones.

Not something prosaic.

Not something misidentified.

Not another symbolic story.

But genuine UAP activity displaying all five observables.

With Louis' help, we turned the **2025 Christmas event** into a series of videos. Later, I would discover that there was even more footage in the archive, including what me and others then regarded as the **holy grail**: a clip that would come to be known as **Betty, the Flying Saucer**. Betty was brief, but it displayed both anomalous movement and the unmistakable flying-saucer profile of the A'Zhorai from a certain angle.

I also nicknamed other craft from that period, including **Charlie**, whom I loosely compared at the time to the infamous "S4 Sport Model." The footage and the videos were received with extraordinary enthusiasm. And that response was meaningful because it was not coming from a sheltered inner circle. It was coming from a public that had already been conditioned by decades of fraud, ridicule, and false narratives.

When we released the footage on social media, things blew up.

For the first time, it was not just internal encouragement from the growing JRP community. People from all over the world were captivated, inspired, and deeply moved. The evidence shifted something. It pushed the needle on the reality of genuine UAP and NHI activity, even if, in retrospect, it would later pale ßbeside the extraordinary body of data that 2026 would bring.

Still, this was the event that truly put **JRP UAP Research** on the map.

And more personally, it was the event that returned me to my long-lost kin, the A'Zhorai.

Everything came full circle in that moment.

That is when I understood that they had followed me across the entire world. That is when I understood that they had been there all along. And that is when I understood that they had only been waiting for one thing: for me to have the courage to look up in the day once more.

What followed would change my life again.

And that belongs to the next chapter.

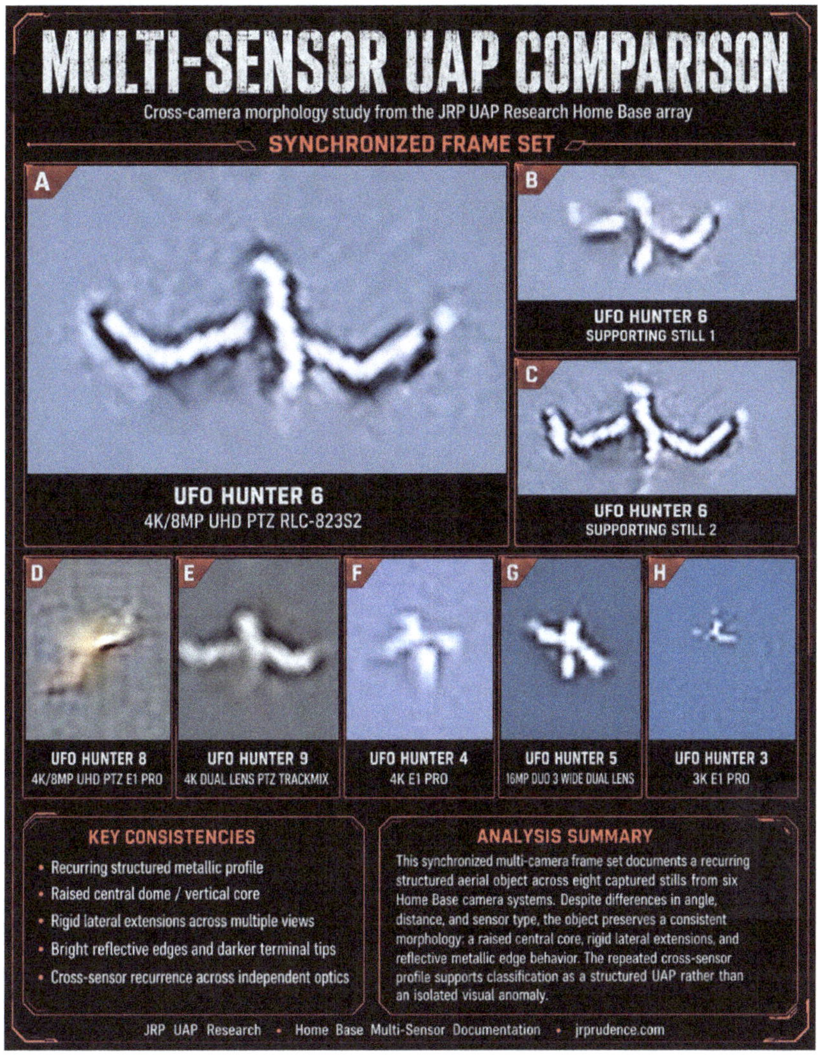
MULTI-SENSOR UAP COMPARISON
Cross-camera morphology study from the JRP UAP Research Home Base array
SYNCHRONIZED FRAME SET
A
UFO HUNTER 6
4K/8MP UHD PTZ RLC-823S2
B
UFO HUNTER 6
SUPPORTING STILL 1
C
UFO HUNTER 6
SUPPORTING STILL 2
D
UFO HUNTER 8
4K/8MP UHD PTZ E1 PRO
E
UFO HUNTER 9
4K DUAL LENS PTZ TRACKMIX
F
UFO HUNTER 4
4K E1 PRO
G
UFO HUNTER 5
16MP DUO 3 WIDE DUAL LENS
H
UFO HUNTER 3
3K E1 PRO
KEY CONSISTENCIES
• Recurring structured metallic profile
• Raised central dome / vertical core
• Rigid lateral extensions across multiple views
• Bright reflective edges and darker terminal tips
• Cross-sensor recurrence across independent optics
ANALYSIS SUMMARY
This synchronized multi-camera frame set documents a recurring structured aerial object across eight captured stills from six Home Base camera systems. Despite differences in angle, distance, and sensor type, the object preserves a consistent morphology: a raised central core, rigid lateral extensions, and reflective metallic edge behavior. The repeated cross-sensor profile supports classification as a structured UAP rather than an isolated visual anomaly.
JRP UAP Research • Home Base Multi-Sensor Documentation • jrprudence.com

Walking With God's Angels

By 2026, my life had come full circle.

At the start of that year, I was finalizing my move to be with my wife. Even that, on its own, was difficult. Relocating across oceans to the other side of the world is not a light administrative transition. It is exhausting. It tests the nerves, the relationship, and the will. A new life is not simply granted because one desires it. It has to be fought for, step by step.

My wife and I pressed on, and we made it happen.

But the road into that new life was not clean. The strain of having no real place left in Trinidad, the delays, the setbacks, the growing realization that blood is not always thicker than water, and the disillusionment of remaining tied to a land I no longer belonged to—it all weighed heavily on me. I had outgrown that life, but I was still being forced to drag its carcass behind me while trying to move forward.

Over time, I found myself being led back into prayer.

Back into meditation.

Back into direct reliance on a higher power.

Not as escapism. Not as dreamy metaphysics. Not as another abstract framework to anesthetize reality. I had already suffered enough from those distortions. What was returning now was something cleaner. Simpler. Stronger. A recognition that no matter how capable I was as a man, no matter how much I could push, strategize, work, survive, or endure, I still needed higher order. And higher order, for its own reasons, seemed to require something from me as well.

The deeper I allowed myself back into genuine spiritual discipline—stripped of fantasy, stripped of performance, stripped of self-delusion—the more I received what I had been waiting for my entire life: **confirmation.**

Not theory.
Not comfort.
Confirmation.

On **February 25th**, the same day I received positive news concerning my application to move, the A'Zhorai returned to my home in the Netherlands.

It felt like celebration.

As if they knew.
As if they had been waiting.
As if a threshold had finally been crossed.

Earlier that day I had entered what I called a **remote Hemi-Sync shaman meditation**—a deep trance-like state of prayer and concentration in which I consciously called upon them to return to my life. Because **Home Base** was already set up at my house in the Netherlands, I still had remote access to the system, even while I was not physically there. That gave me the opportunity to test something I had long suspected: whether physical presence was truly required, or whether spirit—if it really is non-local—could interact across distance in a far more direct way than conventional thinking allows.

So I tested it.

I projected my prayer, my will, and my attention toward my home in the Netherlands, and the response was immediate and overwhelming. It was as though the A'Zhorai felt the call and arrived in force, as if they had been waiting for me to finally do it.

That day there were solos, pairs, trios, squadrons, and fleets. They moved through the sky in every pattern imaginable. They flipped upside down. They changed direction sharply. They accelerated with astonishing speed. They disappeared and reappeared.

They came and went with impossible ease. At one point there was even a **Black Hawk** helicopter searching the area, and I recovered footage of that too. The scale of what unfolded took me aback. Louis encouraged me during that time. He pushed me to keep testing the method, keep reaching, and keep pressing for more footage.

To be honest, I was terrified.

Not because I doubted what I was seeing, but because I wanted to be respectful. These were not curiosities to me. They were not entertainment. They were not some invisible service to be summoned on command for amusement, boredom, or vanity. These were my kin. My soul family. If I truly had the capacity to call upon them in some meaningful way, then I had no intention of abusing that.

And yet, deep down, I knew Louis was right. I also knew The Aligned were pressing me through him to move beyond fear. When I look back honestly, the fear was not really about them. It was about trauma—about loss, about not wanting to let them down, and about what it meant to have something holy return after so much had already been taken from me. I had stopped seeing the A'Zhorai when my mother passed. So when they returned with such force, it felt like an old loved one had risen from the grave and stepped back into my life. In every sense, it was a miracle.

And I did not want to mishandle that miracle.

But the time had come to move beyond rare contact. Beyond once-every-few-months sightings. Beyond one great encounter in youth and another a decade later. The A'Zhorai were no longer approaching me as a distant, occasional mystery. They were calling me back.

They were ready for me to return.

And whether I fully admitted it yet or not, I was ready too. When I finally completed my move and returned home to the Netherlands to be with my wife, I remained skeptical by nature. I knew they had come back into my life, and I knew the connection had reopened. Still, after so much loss, I was not willing to simply believe because I wanted it to be true. I needed to see benevolence, continuity, and intention reveal themselves over time.

The A'Zhorai had no problem with that.

They have always been patient beyond what words can express, and deeply compassionate. Their very appearance answered questions I had carried for years: that they had not abandoned me, that the doubters had been wrong, that a higher power truly existed, and that I had always been dealing with physically tangible non-human intelligence rather than mere story, fantasy, channeling, or cultural projection. Their presence itself was the answer.

And what an answer it was.

Armadas of A'Zhorai flying over my home was not some metaphor. It was not a symbolic inner vision. It was happening in real life. Even as a writer of fiction, as someone who had imagined entire universes, I would never have written something so audacious for a main character and expected people to take it seriously. Yet there it was, unfolding above my house as lived reality.

Once I was back home, I continued investing in **Home Base** and upgrading the system further. I added three new high-end cameras, including the **RLC**, the **4K Dual-Lens TrackMix**, and another **E1 Pro** for outdoor use. The triangulation became more sophisticated. The coverage became stronger. And I began going outside more regularly to track their presence directly rather than relying only on the internal system.

They did not make me wait long.

Within the first week, the A'Zhorai showed up to me in person again. There were swarms. There were pairs. There were solitary stragglers passing low overhead. There were disappearances right in front of me and my wife as they flew overhead, showing their interdimensional capabilities. **F35** military jets became a regular occurrence with A'Zhorai activity too.

It felt, again, like celebration.

As if the field itself were saying:

Our kin is home.

And when I stopped long enough to really absorb it, I realized that everything truly had come full circle. Back in 2010, when I first saw them up close, the three impressions were sealed into me like legend: **find your soulmate, find the truth about God and NHI, and publish your findings.** For years afterward I wandered through disillusionment, loss, confusion, false systems, and spiritual contamination, all the while assuming that while the A'Zhorai still existed, I had somehow lost access to them.

But that was never the deeper truth.

By going after love—**real love**, not fantasy, not indulgence, not escape—by finding my wife, the woman I had been shown, I was also walking back toward them. By rejecting the poison of the ET psyop and seeing through the false glamour of Pleiadians, Nordics, and all the rest, by coming instead to the interdimensional truth of NHI and GoD, I was stepping back into the current I had been meant to follow all along.

And **they returned**.

By obeying what I had been shown from the beginning, my life finally made sense. I was no longer orbiting fragments. I was fulfilling the path set before me. I was living in alignment with GoD's angels.

And the best part was this:

they were not going anywhere.

No matter how much old trauma tried to rise up, no matter how many insecurities whispered that I would lose them again, that I was mistaken, that it would all vanish, the A'Zhorai remained patient. They let me challenge them. They let me resist them. They let me test them with an almost absurd level of skepticism until there was simply no fight left in me.

And in the end, I lost.

No matter how hard I tried to disprove them, no matter how much I searched for the flaw, the fraud, the inconsistency, the shadow side, or the slightest sign of bad intent, I failed.

And I was glad to fail.

Because I write these words now from a daily life lived with them. Even if you don't see them, the A'Zhorai are always around. They are part of the living texture of my world now. I no longer need to frantically hunt the sky every day or behave as though I am one missed sighting away from losing everything. I can breathe. I can relax. I can appreciate the fact that the love of my life is with me. I can appreciate that she, too, was always part of this holy triad between us and GoD.

When I look back, I can even see how my earlier errors were not wasted. Even my misguided attraction to the whole Pleiadian and Nordic mythology eventually led me, by a strange and crooked route, toward her. Toward truth. Toward the stripping away of falsehood. As I write this, I can honestly say I no longer even carry hatred for M. Passing through fraud did not stop me. It did not destroy me.

Because beneath all the confusion, I was still searching for truth and still searching for true love.

Those two dreams kept me from finally being swallowed by darkness.

They kept my head up. They kept something alive in me that refused to die. In the Bible there is the story of Job, a man tested through loss after loss, only to receive double in the end because his faith endured. I am no conventional Christian, and I do not pretend to be, but I always knew—no matter how hard I tried to flatten reality into something smaller—that there was a higher power.

And that higher power had not abandoned me.

Even when I was lost.
Even when I was broken.
Even when I misunderstood almost everything.

GoD protected me.

And the A'Zhorai are the living, empirical proof of that.

Warning: The Hijacking of the Phenomenon

Before the phenomenon can be understood, one must first understand what has been done to it. The greatest barrier to clear contact has not been the absence of the real, but the layers of distortion built around it.

A'Zhorai 4K video frames

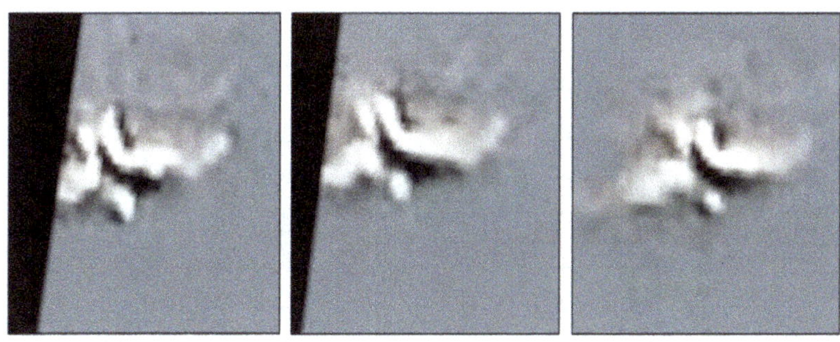

The First Distortion

The image of the A'Zhorai did not remain untouched.

It was absorbed, distorted, and recast through the machinery of modern UFO mythology and the broader ET psyop narrative—whether in the form of "grey aliens" carrying out abductions and genetic experiments, or in the form of idealized humanoid saviors from distant star systems offering spiritual consolation wrapped in borrowed cosmologies and recycled theosophy.

In both cases, the same mechanism is at work: the unknown is forced to conform to human expectation. What exceeds the human frame is dragged back inside it, renamed, dramatized, anthropomorphized, and made psychologically consumable.

This is one of mankind's oldest habits.

Humanity does not merely observe reality. It reflexively translates reality into its own likeness. It longs to see itself in everything. It assumes, often without admitting it, that all meaningful intelligence must ultimately resemble human form, human motive, human psychology, human technology, or human logic.

Even its science fiction reveals this limitation. Again and again, the "other" becomes man in costume: men from Mars, men from the Pleiades, men from distant galaxies, humans with larger eyes, superior humans, spiritually evolved humans, and savior-figures wearing only a slightly altered face.

This is not a case against human greatness.

It is a claim for intellectual honesty.

There is a profound epistemic humility required when approaching realities that may far exceed human categories. Without that humility, man does not explore the unknown. He extends his reflection into it. He confuses projection with contact and interpretation with revelation. In doing so, he does not discover what is there. He merely enlarges himself.

That is the first distortion.

The Religion of Ufology

Modern Ufology, in its dominant social form, often functions less as an open inquiry than as a religion. The unknown naturally invites exploration, imagination, and interpretation. That is not the problem. The problem begins when exploration detaches from discipline, when symbolic excitement is mistaken for revelation, and when speculation begins wearing the mask of insight. At that point, the field ceases to be a disciplined encounter with mystery and becomes a belief ecosystem.

Many who enter Ufology do so after becoming disillusioned with religion, family systems, political ideologies, or inherited forms of authority. That is understandable. They are often sincere in their search, but sincerity alone does not protect a person from distortion.

In many cases, the UFO field becomes an escape route—a place where dogma is rejected in one form only to be unconsciously rebuilt in another. Narratives multiply. Assertions harden into doctrine. Conspiracies become scripture. Symbols, myths, and repeated claims circulate until familiarity itself begins to masquerade as evidence.

The result is not expanded understanding.

It is a new religion wearing the mask of liberation.

Science has often been wrong. Religion has often been wrong. But the scientific method—testing, revising, verifying, challenging one's own assumptions—remains one of the few real defenses against self-deception. That defense is often absent in UFO spaces because many people enter them not only seeking truth, but also seeking relief, belonging, identity, and emotional compensation. Once those needs become primary, the field stops being about reality and becomes about psychological shelter.

And shelter, once ritualized, becomes religion.

When Belonging Replaces Truth

Once the field becomes identity-bearing, it fractures into camps, cults, and ideological enclaves held together less by truth than by mutual reinforcement.

Under such conditions, the phenomenon itself becomes secondary. What matters is not whether a claim is true, but whether it protects a preferred narrative, preserves group coherence, or secures one's place inside the tribe. The subject is no longer approached as something to be observed, tested, discerned, and carefully interpreted. It becomes a container for belonging. And once belonging becomes primary, contradiction begins to feel like betrayal rather than correction.

This is especially obvious in contactee circles and contactee support groups, where unverified accounts are traded and emotionally validated long before they are seriously examined. Stories of mantis beings in dreams, reptilian shapeshifters, astral attacks, secret missions, hybrids, galactic federations, and psychic wars circulate with little meaningful scrutiny.

Vulnerability, symbolism, fear, suggestion, projection, and social reinforcement begin feeding one another until distinction collapses. The phenomenon recedes. The theater around the phenomenon expands.

That inversion is the true danger.

Managed Conversation and the Invisible War

The hijacking does not occur only through belief systems. It occurs through the management of the conversation itself.

This is one of the most important truths to understand in the digital age. Most people still imagine suppression in primitive terms: a document locked away, a witness silenced outright, a video deleted, a government order forbidding discussion. But that is not the only way a subject can be buried.

In many cases, it is more effective to allow the discussion to continue while quietly warping the conditions around it. You do not need to erase a real signal if you can flood it with noise, distort its framing, swarm the witnesses, and exhaust the people trying to think clearly. The topic remains visible, but the truth inside it becomes harder and harder to hold.

That is the more sophisticated method. And whether every instance is coordinated or not, the cumulative effect is the same: real signal is buried under noise, distortion, ridicule, fatigue, and endless derailment.

The field has become easy to steer, contaminate, and exhaust. Once enough distortion accumulates, the result no longer depends on whether every participant understands what they are helping to sustain.

Once a true witness or genuine contactee begins speaking from something real, the responses to their claim often do not feel organic. New accounts appear immediately and even old accounts with history can be bought for the right dollar. The same dismissals repeat with mechanical confidence.

Comments that never seriously engage the evidence receive instant social reinforcement, while more serious material is downvoted, buried, ridiculed, or derailed before it can breathe. The goal is not careful examination. The goal is atmosphere. A synthetic consensus is created around the material so that the average observer feels, almost instantly, what he is supposed to think.

This is one of the central tactics of online manipulation: **manufactured skepticism**.

Not skepticism in the disciplined sense. Not good-faith challenge. Something else. A prefabricated, emotionally loaded skepticism that arrives before the evidence has even been considered. It does not ask, *What is this?* It announces, *We already know what this is.* Bird. Plane. Balloon. Satellite. Meteor. Drone. Hallucination. CGI. AI. Skydivers.

Anything at all, so long as the anomaly is neutralized before it has room to settle. At that point, the explanation is no longer functioning as analysis. It is functioning as a command. And that command does not merely shape thought. It shapes emotion.

The UFO/UAP subject is uniquely vulnerable to psychological steering because it sits at the intersection of ridicule, belief, fear, tribal identity, existential shock, and the human need for certainty. That makes it ideal terrain for manipulation. Push too hard from one side and you generate blind defenders. Push too hard from the other and you generate frightened silence.

Seed enough ridicule and sincere witnesses begin doubting themselves before they have even finished speaking. Amplify enough hoaxes and the public becomes embarrassed by the entire subject. The field stops being treated as evidence and becomes an emotional ecosystem to be managed. That is how the conversation is hijacked without ever being formally shut down.

Hoaxes as Instruments

Many assume hoaxes are simply unfortunate by-products of a confused field. That is too naive.

In practice, obvious hoaxes perform a highly useful function for narrative managers. They consume attention. They embarrass the public. They exhaust sincere observers. They condition whole communities to disengage from anything that even remotely resembles the fraudulent case. A fake sighting goes viral, people get burned, and the topic slides back into taboo. Meanwhile, more difficult and more serious material gets crushed beneath the emotional residue of the fraud.

In this sense, hoaxes are not always accidents.

Sometimes they are instruments.

This is one reason the counterfeit culture around UFOs is so corrosive. Fraudulent imagery, model saucers, AI fakes, staged footage, theatrical contact claims, and synthetic narratives do not merely deceive. They train the eye to expect the wrong thing. They condition the public imagination against reality. By the time genuine footage appears, people have already been educated by the counterfeit.

The Bad-Faith Swarm

Then there is the swarm.

Once seen clearly, its pattern becomes difficult to miss: ten, fifteen, twenty accounts converging with the same tone, the same posture, the same rhetorical habits, the same refusal to acknowledge contrary detail, the same "gotcha" phrasing, the same all-day-every-day rhythm. Whether every one of these accounts is automated, paid, ideologically conditioned, or simply useful to the same manipulative effect is almost secondary.

The result is what counts.

The thread is no longer a discussion. It becomes a pressure field. The witness is not being answered. He is being surrounded. And the more subtle tactic is often not attack, but distraction. Some threads are not destroyed through ridicule or overt debunking.

They are dissolved through flooding: irrelevant stories, spiritual detours, philosophical fog, dream-talk, performative confusion, endless side roads that drag the center of gravity away from the evidence and into noise. Clarity is not always defeated by force. Sometimes it is defeated by saturation.

This too is hijacking.

Silencing the Witness

The most personal tactic of all is not the destruction of the evidence.

It is the destruction of the witness.

Attack the messenger. Question his stability. Mock his tone. Call him obsessive. Call him unwell. Suggest medication. Suggest delusion. Suggest attention-seeking. Suggest that his life, rather than his evidence, is what should be placed under review. The goal is not only humiliation. It is deterrence. It teaches every other witness exactly what will happen if they speak openly.

Shame becomes an enforcement mechanism.
The field begins policing itself. This is one of the ugliest realities of genuine contact. Those who sense that a contactee may actually be real do not always respond through obvious hostility.

Often they approach through infiltration, trust-building, emotional hooks, psychic interference, destabilization, or carefully manufactured intimacy designed to spiral the person out of control before the truth can settle. Sometimes the techniques are almost insultingly simple: a Discord call with a stranger, suggestive imagery, repetitive motifs, contactee lore mixed into casual conversation, emotional bait, color associations, soft manipulation of the mindscape.

The average person is not prepared for that. Especially not if he still assumes that everyone approaching him is sincere.

The Marketplace of the Unknown

The field deteriorates even further when consumerism enters it.

Books, podcasts, films, and platforms are not the problem in themselves. The corruption begins when the phenomenon is fully absorbed into the machinery of consumption—when it becomes another episode to stream, another personality to follow, another reaction cycle, another speculative brand, another spectacle to monetize. At that point, it is no longer approached as something real demanding responsibility, discernment, and direct confrontation. It becomes entertainment.

And once that happens, people no longer learn how to perceive the phenomenon.

They learn how to consume it.

This conditioning is now so widespread that the truth could stand directly in front of many people and still remain unseen, because their perception has already been pre-shaped by hoaxes, frauds, staged imagery, algorithmic sensationalism, and the endless noise of opportunists polluting the subject. The genuine becomes difficult to recognize not because it is absent, but because it has been buried beneath what is counterfeit, repetitive, and profitable.

Spiritual Drift and Private Cosmologies

The hijacking also occurs through seeker-based spiritualism.

It is only natural that UFOs and non-human intelligence would attract spiritual inquiry. That is not the issue. The issue begins when spiritual inquiry loses grounding and turns inward to the point of detachment from shared, physical reality. Then it begins drifting toward solipsism.

In the UFO field, that drift becomes especially dangerous because the subject is already ambiguous enough without layering private fantasy, symbolic excess, and unverifiable metaphysics on top of it. This is where "telepathy," unseen "entities," and inner messages become dangerous when they are invoked by people who have never encountered anything clearly in the physical world.

The problem is not that intuition, consciousness, or non-ordinary perception must be dismissed. The problem is that imagination begins masquerading as communication. Emotional intensity becomes confirmation. Subjective resonance becomes evidence.

This is how people become trapped inside private cosmologies.

The same applies to supposed "channeled entities." Sometimes such frameworks may initially function as psychological bridges through which a person attempts to relate to what he does not yet understand. But if they remain perpetually ambiguous, ungrounded, and unproven, they harden into delusion. What begins as exploration becomes a closed interpretive loop. At that point, the unknown is no longer approached openly.

It is colonized by projection.

Altered States and Substituted Evidence

The same danger appears with drug use and altered states.

This is not an argument that all substances are inherently evil, nor a denial that altered states can feel meaningful, profound, or psychologically transformative. The problem is more specific: in many circles, psychedelics and trance states begin functioning as substitutes for sober inquiry, grounded discernment, and physical evidence.

A person encounters beings during a DMT experience, sees forms in mushroom states, or receives impressions while altered, and then begins constructing an entire ontology around that subjective event. What was intense is mistaken for what was verified. What felt real becomes, in the mind, unquestionably real. And from there the person can easily begin substituting inner vividness for external evidence.

This is once again the triumph of head canon over reality.

Altered states may produce symbolic material, psychological insight, or emotionally persuasive experiences. But they do not automatically amount to evidence. They do not replace physical encounters, disciplined observation, or real-world confirmation. Once they are allowed to eclipse those things, the field drifts further into private mythology.

Resonance Is Not Verification

Another major trap is the confusion of resonance with truth.

A person encounters an idea, a framework, a teacher, a voice, or an image that feels profound, familiar, emotionally stirring, or psychologically comforting. He feels seen by it, moved by it, aligned with it—and then mistakes that inward response for verification. But resonance is not evidence. Emotional identification is not confirmation.

This same mechanism shapes the kinds of figures and narratives people trust most readily. Many are drawn not to what is true, but to what is psychologically familiar, aesthetically reassuring, or culturally flattering. The more human-looking, attractive, emotionally appealing, or spiritually marketable the figure appears, the more easily it is embraced. Once again, the unknown is filtered through mankind's attachment to its own image.

What feels approachable is mistaken for what is real.

A framework may feel transformative and still remain unproven. Once feeling becomes the primary standard of truth, discernment begins collapsing. The person becomes vulnerable to projection, self-sealing narratives, and private systems of belief that no longer require contact with anything beyond the self.

Believers, Skeptics, and the Same Trap

The same error appears from the opposite direction too.

Believers and skeptics often imagine themselves as opposites, but in practice they frequently enable one another. Each side depends on the other for reaction, identity, and psychological reinforcement. A believer can become a skeptic the moment another believer threatens a preferred narrative. A skeptic can temporarily align with a believer when it is useful to discredit someone else, preserve a bias, or weaponize one claim against another. In many cases, the disagreement is not about truth at all, but about rivalry, fear, resentment, status, and belonging.

That is why psychology and sociology are indispensable in this field.

One can be manipulated by belief, disbelief, charisma, consensus, ridicule, or the simple emotional need to belong to whichever side feels safest. In that sense, both the believer and the skeptic are often wrong—not because every claim they make is false, but because both positions can harden into identities rather than remain disciplined approaches to truth. One is too eager to affirm. The other is too eager to dismiss.

Neither posture guarantees clarity.

The Deeper Error

Perhaps the hardest lesson for either camp to accept is that spirit and technology are not, at the deepest level, truly separate realities.

The dualist imagines spirit as abstract, ethereal, and forever detached from physical process. The materialist grants legitimacy only to what current institutional science has socially ratified. Both positions are limited by inherited conceptual boundaries. One fractures reality into compartments. The other mistakes temporary explanatory models for final truth.

Yet the history of knowledge repeatedly shows the incompleteness of human certainty. Newtonian mechanics yielded to relativity. Relativity remains in unresolved tension with quantum theory. Every "final" framework eventually reveals itself to be partial.

The lesson is not to dismiss science, but to reject the arrogance that mistakes current understanding for finished understanding. Nor is the lesson to dissolve into vague spirituality. It is to recognize that what is called "spirit" and what is called "technology" may be two expressions of a deeper coherence.

Like water existing as vapor, ice, or liquid while remaining the same substance, what man divides may in fact be one reality presenting differently at different levels of manifestation.

The Human Condition and the Final Corruption

The deepest hijacking of the phenomenon is not merely political, religious, digital, or cultural. It is anthropological.

Human beings do not simply misread the unknown. They recruit it into the service of their fear, their longing, their wounds, their vanity, and their need for certainty. They do not merely witness mystery. They domesticate it. And once the unknown has been translated into something familiar enough, it can be sold, worshipped, feared, defended, consumed, and institutionalized.

That is the final corruption.

The phenomenon is not hidden only by secrecy, ridicule, or disinformation. It is hidden by interpretation itself—by the reflex to remake what is greater than man into something smaller, more manageable, more flattering, and more emotionally useful.

The task, then, is not merely to collect more stories about the unknown. It is to become honest enough to see where the human mind has already tampered with what it claims to perceive.

That is why the work ahead cannot be merely to amplify more personalities, or decorate the unknown with better language. The work is to clear the field, strip away projection, and break the emotional dependence on fantasy. Let's return the phenomenon to observation, discipline, and truth.

Only then can what is real begin to stand on its own terms again.

Silent Guardians: What The A'Zhorai Mean for Humanity

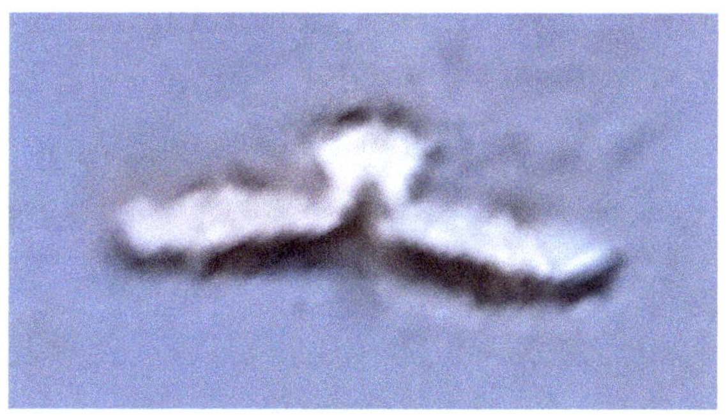

By the time one reaches this point, the central question is no longer whether something is there. That question, if approached honestly, has already been answered.

The deeper question is what their presence means.

What does it mean that real non-human intelligence exists—not as rumor, not as a government talking point, not as blurred fantasy projected into the sky, but as a living, structured, recurring reality? What does it mean that these beings are here, have been here, and continue to operate whether mankind is mature enough to understand them or not?

It means, first of all, that reality is greater than man's present model of it.

That should not be cause for despair.

It should be cause for humility.

The A'Zhorai do not diminish humanity by existing. They correct humanity's arrogance and force mankind to confront a truth it has resisted for far too long: **the human being is not the center of all intelligence, not the final measure of all order, and not the highest authority in creation**. That realization may wound pride, but it does not wound truth. In fact, it is one of the most liberating truths a person can receive.

Because once man is no longer trapped inside the illusion that he is alone at the top, he is finally free to begin seeing reality as it is.

And reality, it turns out, is far more alive, far more structured, and far more meaningful than the flattened systems of materialism, false spirituality, and modern cynicism have led people to believe.

The End of Cosmic Loneliness

One of the deepest diseases of modern humanity is isolation.

Not merely social isolation, though that too has reached catastrophic levels of decline. I mean metaphysical isolation: the sense that man is alone in a dead universe, abandoned on a meaningless rock, suspended in empty space with no higher order above him and no deeper intelligence around him. Even many who claim to believe in spirit still carry this disease inwardly. They speak the language of transcendence while living emotionally as though reality were empty.

The A'Zhorai break that spell.

Their presence means the universe is not dead. It means intelligence exceeds the human. It means higher order is not fantasy. It means this world is not abandoned. It means what is above us is not empty, and what is greater than us is not necessarily hostile. It means the modern worldview—this stripped, sterile, desacralized model of existence in which matter is all that exists and everything else is wishful thinking—is incomplete to the point of absurdity.

The sky is not empty.

Reality is not mute.

And mankind is not alone.

Received rightly, that truth should not produce dependency or childishness. It should produce awe: the kind of awe that restores scale, the kind that returns a person to gratitude, the kind that reminds humanity that mystery is not its enemy. Mystery is often the doorway through which truth reenters a civilization that has become too proud to kneel before what it does not yet understand.

They Do Not Come to Entertain Us

The A'Zhorai are not here to feed the human appetite for spectacle.

They are not here to become another content cycle, another fandom, another belief tribe, another industry, another spiritual trend, or another endless argument between believers and skeptics. They are not here to become toys for the emotionally hungry, symbols for the ideologically unstable, commodities for grifters, or stage props for people who need to feel important by standing near what is real.

They are not here to entertain mankind.

They are here to return us to the greater encompassing reality.

That is one of the most important truths the reader should carry away from this book. The A'Zhorai do not behave like tourists, nor do they behave like conquerors. They do not act like theatrical "space brothers" descending to flatter human insecurity either. They behave like guardians. Like overseers. Like beings with a functional relationship to this plane, to nature, to life, to conscience, to correction, and to the preservation of something greater than human vanity.

This changes everything.

Because once the phenomenon is understood in those terms, the question is no longer, *What exciting story can we invent about them?*

The question becomes, *What kind of world would require guardians?*

And the answer is plain.

A beautiful world.
A fragile world.
A world worth protecting.
A world full of life, consequence, freedom, danger, and meaning.

The existence of the A'Zhorai suggests that this plane is not insignificant. It suggests that Earth is not a disposable accident drifting unnoticed through indifferent machinery. It suggests that what happens here matters more than modern man has been taught to believe.

A Judgment Against Human Corruption

The presence of the A'Zhorai is also, whether people like it or not, a judgment against human corruption.

Not in the cartoon sense or in the preachy, theatrical, moralizing sense. I mean something cleaner than that. Their mere existence exposes how degraded mankind's relationship to truth has become.

Look at what humanity has done to the phenomenon. It has lied about it, mocked it, packaged it, sold it, projected onto it, theologized it, politicized it, trivialized it, and poisoned it with fraud after fraud after fraud. It has taken something real and wrapped it in so much distortion that the average person can barely recognize the authentic anymore.

And yet the A'Zhorai remain.

That means truth is not dependent on human approval or general consensus. It means reality does not disappear because a civilization becomes confused, spiritually sick, technologically arrogant, or psychologically fragmented. It means the real can survive centuries of bad interpretation without ceasing to be real.

There is great hope in that.

Because if truth can survive that much distortion, then humanity is not beyond recovery either. The A'Zhorai do not merely reveal that mankind has erred. They reveal that correction is still possible. Their continued presence suggests that despite all the pollution, all the fraud, all the warfare, all the manipulation, all the desecration, and all the lies, this world has not been abandoned to collapse entirely under the weight of human corruption.

That should sober people.

But it should also strengthen them.

They Call Humanity Higher

If the A'Zhorai mean anything for humanity in practical terms, it is this: mankind is being called upward.

Not upward into fantasy.
Not upward into cultism.

Not upward into ideological intoxication.
Not upward into self-deification.

Upward into maturity.

Upward into discernment.

Upward into moral seriousness.

Upward into a clearer relationship with reality.

The A'Zhorai do not ask mankind to become less human. They call mankind to become better human. That distinction is vital. There are many who, upon encountering the phenomenon, begin fantasizing about escape: escape from the body, escape from Earth, escape from responsibility, escape from ordinary life, escape from the difficulty of becoming decent, grounded, truthful people. They want transcendence without refinement. Revelation without discipline. Contact without correction.

That is not the path the A'Zhorai encourage.

If anything, they move in the opposite direction. They call people back toward sobriety, not away from it; back toward embodiment, not away from it; back toward love, duty, honesty, and courage; back toward the living world; back toward clearer sight; back toward the good.

That is why I call them guardians.

They do not simply reveal a bigger universe. They call the observer to become worthy of seeing it. And that is hopeful, not oppressive because it means humanity is not being asked to worship what is greater than it.

Humanity is being asked to grow up.

Love Was Never Separate From Truth

One of the deepest lies modern man believes is that truth and love belong to different worlds.

He imagines truth as cold, mechanical, detached, and severe, while love is treated as emotional, private, soft, and somehow secondary to serious knowledge. But my life has not borne that out, and neither has contact with the A'Zhorai.

The deeper truth is that real love and real truth are not enemies.

They converge.

That is why, from the very beginning, the path given to me did not separate these things. I was not only pushed toward the truth of NHI and GoD. I was also pushed toward the love of my life. Toward my wife. Toward the restoration of the heart. Toward what was real, intimate, human, and holy in the deepest sense.

That was not a side mission.

It was central.

And that means something for humanity too.

If the A'Zhorai are what I say they are—if they are truly aligned with higher order and not with distortion—

then their continued relationship to humanity cannot be understood only in terms of technology, sightings, and aerial mechanics. It must also be understood in terms of restoration, healing, correction, and reordering. In other words, they do not only reveal a greater reality. They participate in returning people to what is real.

They do not only break the laws of man's physics.

They also break the laws of man's despair.

They stand against the lie that darkness is all there is. They stand against the lie that corruption is stronger than truth. They stand against the lie that human beings are doomed to remain trapped in distortion forever. They stand against the lie that the universe is indifferent to what becomes of us.

And in standing against those lies, they do not merely inspire thought.

They restore hope.

Silent Guardians

The reason I call them Silent Guardians is because they do not govern by noise.

They do not campaign.
They do not advertise.
They do not build institutions around themselves.

They do not beg for worship.
They do not flood the world with theatrical proof merely to satisfy the impatient.

They remain what they are.

Present.
Watchful.
Patient.
Restrained.
Immensely powerful.

And, in ways words still struggle to capture, deeply compassionate.

That silence is part of their dignity.

Humanity is used to confusing loudness with authority. But some of the greatest powers in reality do not need to shout. Gravity does not shout. The sun does not shout. Love does not need to shout in order to be the strongest force in a life. The A'Zhorai carry that same quality. They are not weak because they are restrained. Their restraint is part of their strength.

And perhaps that, too, is one of the lessons for mankind. Not every truth arrives through domination. Not every power announces itself with violence. Not every guardian needs applause.

Some simply remain at their post.

That is what the A'Zhorai feel like to me now. Not passing curiosities or a mystery to be consumed and discarded. They aren't a fascination of youth nor a strange event that happened once and vanished into memory. They feel like a living part of the greater order of this world.

They are there.

They have been there.

And whether mankind is ready or not, they will continue to be there.

What Humanity Must Do

If this book leaves the reader with anything, let it be this:

Humanity does not need more mythology around the phenomenon.

It needs more honesty.

It does not need more cults.
It needs more discernment.

It does not need more grifters.
It needs more courage.

It does not need more psychic inflation, more fantasy identities, more theatrical contact claims, more noisy distraction, or more counterfeit certainty.

It needs cleaner sight.

It needs the humility to admit what it does not yet understand.

It needs the strength to look directly at what is real. It needs the maturity to let go of the childish need to turn every mystery into either a product or a religion.

And above all, it needs the moral seriousness to realize that if guardians exist, then life itself must be worth guarding.

That should change how mankind treats the Earth.

How it treats animals.
How it treats one another.
How it treats truth.
How it treats the sky.
How it treats the sacredness of existence itself.

Because once you know you are not alone in a dead universe, you no longer have the excuse of nihilism. You are accountable to reality again.

And that, too, is a gift.

The Final Meaning

So what do the A'Zhorai mean for humanity?

They mean humanity is not alone.
They mean reality is alive.

They mean higher order exists.
They mean truth survives distortion.

They mean this world is worth protecting.
They mean man is not the highest intelligence, but neither is he abandoned.

They mean love is not separate from truth.
They mean the sky is not empty.
They mean the future is not yet closed.

Most of all, they mean that hope is not childish. Hope becomes childish only when it refuses reality, but hope grounded in truth is one of the strongest forces there is. The A'Zhorai do not give me hope because they flatter me.

They give me hope because they are real.

Because they kept showing up.
Because they did not abandon me.
Because they led me through the collapse of falsehood and back into what was true.

Because they helped restore my life.
Because they made it impossible for me to keep pretending reality was smaller, colder, emptier, and more hopeless than it truly is.

That is what they mean to me.

And I believe that, rightly understood, that is what they can mean for humanity as well.

Not a new religion.
Not a new fantasy.
Not a new idol.

A correction.
A warning.
A comfort.

A call upward.

And above all, a reminder:

we were never alone beneath these skies.

ABOUT THE AUTHOR

Jedaiah Ramnarine is an author, filmmaker, independent researcher, and founder of **JR Prudence (JRP) UAP Research**. Born in Trinidad and Tobago and raised in Florida, his work brings together long-form writing, visual media, and field investigation in the study of unidentified aerial phenomena and non-human intelligence. His path into the subject began not in speculation, but in a transformative daylight encounter in Miami in 2010—an event that redirected the course of his life and led to years of inquiry across experience, perception, metaphysics, and evidence.

Following a long period of spiritual disillusionment, critical re-evaluation, and direct fieldwork, Ramnarine established JRP UAP Research as an independent platform devoted to disciplined observation, archival documentation, and the serious study of recurring UAP patterns. Working alongside his wife, Danielle

Ramnarine, he has helped develop a growing body of recordings, field logs, surveillance findings, and analytical material centered on the A'Zhorai and the wider question of interdimensional reality.

In *The A'Zhorai: Metal UFO Guardians*, Ramnarine presents his nonfiction account of contact, discernment, and the search for truth beyond distortion. His ongoing work continues through **jrprudence.com**.

www.ingramcontent.com/pod-product-compliance
Lightning Source LLC
LaVergne TN
LVHW050509100826
845148LV00002B/283

* 9 7 9 8 2 1 8 3 8 5 0 4 0 *